Annals of Mathematics Studies

Number 37

Lectures on
the Theory of Games

Harold W. Kuhn

PRINCETON UNIVERSITY PRESS

PRINCETON AND OXFORD

2003

Library of Congress Cataloging-in-Publication Data

Kuhn, Harold W. (Harold William), 1925–
Lectures on the theory of games / H.W. Kuhn
p. cm.—(Annals of mathematics studies ; no. 37)
Includes biographical references and index.

ISBN: 978-0-691-02772-2

1. Game theory. I. Title. II. Series
QA269 .K83 2003
519.3–dc21   2002066294

British Library Cataloging-in-Publication Data is available.

Originally issued as a report of the Logistics Research Project
sponsored by the Office of Naval Research,
Contract N6onr-27011, Project No. NR 047-002

This book has been composed in Times Roman
Printed on acid-free paper.

www.pupress.princeton.edu

Printed in the United States of America

3 5 7 9 10 8 6 4 2

Lectures on
the Theory of Games

Harold W. Kuhn

PRINCETON UNIVERSITY PRESS

PRINCETON AND OXFORD

2003

Library of Congress Cataloging-in-Publication Data

Kuhn, Harold W. (Harold William), 1925–
Lectures on the theory of games / H.W. Kuhn
p. cm.—(Annals of mathematics studies ; no. 37)
Includes biographical references and index.

ISBN: 978-0-691-02772-2

1. Game theory. I. Title. II. Series
QA269 .K83 2003
519.3–dc21   2002066294

British Library Cataloging-in-Publication Data is available.

Originally issued as a report of the Logistics Research Project
sponsored by the Office of Naval Research,
Contract N6onr-27011, Project No. NR 047-002

This book has been composed in Times Roman
Printed on acid-free paper.

www.pupress.princeton.edu

Printed in the United States of America

3 5 7 9 10 8 6 4 2

Contents

Author's Note

The reader is deserved an explanation as to why these lectures are published nearly 50 years after they were taught as a course in the mathematics department at Princeton University. The text was submitted, and you could read in the Annals Studies of that time: "Annals Study 37, in press, $3.00." At that time, I withdrew the manuscript for alterations, primarily hoping to add something on the rapidly developing theory of n-person games. The revisions were never made, and the lectures were never published.

This year, Vickie Kearn joined the Princeton University Press and initiated a project to put all of the Annals Studies in print ("on demand"). This is a project that I have supported both in theory and in practice (providing copies of rare Annals Studies which even the Princeton Press did not have). It seemed only appropriate to close the gap caused by the absence of Annals Study 37 in the sequence. In fact, there exist thousands of copies of these lectures throughout the world. They were issued as a project report and I have personally made many photocopies in response to requests.

In short, here are the lectures as they were given to a group of upper level students in the spring of 1952, possibly the first course in game theory taught in any university. There are a number of firsts here, such as the introduction of the term "matrix games."

I have been helped by Gerree P. Pecht, Technical Typesetting Specialist, Department of Mathematics, Princeton University, who has reproduced the original style (typed by IBM typewriter in 1953) with skill and devotion. Her professionalism has been beyond description. In addition, I am grateful to Linny Schenck and Alison Anderson for their careful copyediting.

Harold W. Kuhn

Princeton, New Jersey
April 2002

Preface

These lectures were presented to a group of upper division and graduate students at Princeton University in the spring term of 1952. They are intended as an introduction to the relatively modern mathematical discipline known as the Theory of Games. In an attempt to make them somewhat self-contained, considerable space has been devoted to topics which, while not strictly the subject matter of game theory, are firmly bound to it. Principally, these are taken from the geometry of convex sets and the theory of probability distributions. No previous knowledge of these subjects has been assumed; it is only supposed that the reader has an acquaintance with the calculus and with elementary matrix theory and vector analysis. References to tangential matters are given in notes appended to each chapter.

With a few exceptions, the text follows the lectures exactly as they were presented; the author is indebted to Mr. C. S. Coleman for keeping an accurate record of the content. There has been one deliberate and significant omission, namely, the theory of games with more than two participants. It was felt that this portion of the theory is in such a violent conceptual transition that it would require more time and space for an adequate synthesis than these lectures provided.

Valuable contributions have been made by a number of individuals. In particular, thanks are due to L. S. Shapley and I. Glicksberg for fruitful discussions of the presentation of various sections; to David Gale and A. W. Tucker, with whom the author has developed his attitudes toward the subject matter of game theory; to L. Hutchinson, who found a number of errors; and to Euthie Anthony, who did the painstaking work of typing the master copy of the original report of the Logistics Research Project.

Harold W. Kuhn

Bryn Mawr College
September 1953

Chapter One

What Is the Theory of Games?

The most casual observer of the social behavior of mathematicians will remark that the solitary mathematician is a puzzle-solver while mathematicians together consume an extraordinary amount of time playing games. Coupling this observation with the obvious economic interest of many games of skill and chance, it does not seem strange that the basis for a mathematical theory of games of strategy was created by the mathematician John von Neumann [1]* and elaborated with the economist Oskar Morgenstern [2]. In this series of lectures we shall attempt to present the salient results of this theory.

Although our primary concern will be with the mathematical aspects, a complete understanding of the theory can only come from the recognition that it is an interpreted system. This means that, like any satisfactory physical theory, it consists of (1) an axiom system and (2) a set of semantical rules for its interpretation, that is, a set of prescriptions that connect the mathematical symbols of the axioms with the objects of our experience. In the immediate interpretation, these objects are ordinary parlor games such as chess, poker, or bridge, and the success of the axiom system that we choose is measured by the number of interesting facts that we can derive about these games. There are, of course, other interpretations of the axioms, and one of the more ambitious hopes of the theory is that eventually these will include a substantial portion of the science of economics [3]. In these lectures we will restrict ourselves, with few exceptions, to the principal interpretation in terms of games.

Ideally, such an interpreted system could be presented by giving its axioms and the rules for their interpretation and then deriving all the important theorems of the theory by the application of mathematical techniques. But practically the situation is such that most fields of science seem at the present time to be not yet developed to a degree which would suggest this strict form of presentation; only certain branches of physics and geometry qualify by their apparent completeness. Fortunately for the researcher, the theory of games is by no means complete. Indeed, in some portions, such as games played by more than two players, it seems likely that the final formulation will be far removed from the present theory. Accordingly, the ideal order of presentation will be reversed in these lectures, and it will only be after a number of specific games have been examined in detail that the axioms of the theory will appear.

---

*Numbers in square brackets refer to explanatory material appended to each chapter. In some cases this material leads to sources and advanced topics in the theory.

The central problem of game theory was posed by von Neumann as early as 1926 in Göttingen. It is the following:

If n players, $P_1, \ldots, P_n$, play a given game Γ, how must the ith player, P_i, play to achieve the most favorable result for himself?

Stated in this form, of course, the question is very vague and our first problem will be to sharpen the statement until we can agree upon a meaning. First of all, we mean by a game any of the ordinary objects designated by this name, for instance, chess, checkers, tic-tac-toe, gin rummy. However, unfortunately, the word "game" is used with two meanings in everyday English. We speak of "the game of checkers," referring to the game defined by its rules, and we say that we played "five games of checkers," meaning that we completed five particular contests of checkers. We shall separate these meanings, reserving *game* for the first meaning only and using the word *play* for an individual contest. The usage in French is more fortuitous than in English; *un jeu* is used in the sense that we have just defined for "game" while *une partie* is a play. A similar distinction must be made between the occasion of a selection of one of several alternatives, which we shall call a *move*, and the particular selection, which we will call a *choice*. Thus, when the record of a chess game begins:

1. $P-K4$
2. $P-K4$
3. $Kt-KB3$

the moves are 1, 2, and 3 while the choices are $P-K4$, $P-K4$, and $Kt-KB3$.

In order to give meaning to the phrase, "the most favorable result," we shall assume that, at the end of each play of Γ, each of the players, P_i, will receive an amount of money, h_i, called the *payoff to player P_i*, and that the only object of a player will be to maximize the amount of money he receives [4]. In most parlor games, the payoffs satisfy the condition

$$(1) \qquad h_1 + h_2 + \cdots + h_n = 0 \text{ for all plays.}$$

This important category of games will be called the *zero-sum* games.

Thus we find that, by merely introducing this terminology, we can classify games according to the number of players who participate, the number of moves in the game, and whether the game is zero-sum or not. There is one other immediate property that can be used for classification. All parlor games have the characteristic property that, at each move, only a finite number of alternatives are presented to a player. Also, the "length" of a play or the number of possible choices occurring in a play is bounded by a "stop rule." This rule is simple in some cases (in tic-tac-toe, the maximum number of choices is nine, the number of blank squares in the game diagram), while in other games it is quite complicated (in chess, the length of a play is made finite by rules preventing repeated cycles of choices). Such games will be called *finite* games. However, there certainly exist activities that seem to be games in which people are presented with an infinite number of alternatives, and one can also conceive of games in which a play would consist of an infinite sequence of choices. An example of the first type is provided by a duel in which the duelist has the alternatives of firing at any instant in a fixed interval of the infinite time continuum, while it is a simple matter to construct conceptual games of the second type. Such games will be called *infinite* games.

The preceding classification suggests that we should attack the central problem posed above for the simplest type of game in this hierarchy. Leaving the case of one player aside for the moment, this simplest game must have two players, two moves, each with a finite number of alternatives, and be zero-sum. It is this class of games that we shall analyze in the next chapter.

Notes

1. The birth of the theory of games (or perhaps the birth announcement) was the paper of von Neumann, "Zur Theorie der Gesellshaftsspiele," *Math. Annalen,* **100** (1928), 295–300. This paper contains the min-max theorem for zero-sum two-person games and the core of the characteristic function treatment of games with three or more players.

2. The infant of 1929 appears full grown in the impressive volume of von Neumann and Morgenstern, *Theory of Games and Economic Behavior*, Princeton, 1944, 2nd ed. 1947. Here we find the economic interpretation given full justice.

3. The immediate hope inspired by the appearance of the book is reflected in the reviews in economic and mathematical journals. The following might be mentioned as being representative:

Hurwicz, L., "The theory of economic behavior," *American Economic Review*, **35** (1945), 909–925.

Marschak, J., "Neumann's and Morgenstern's new approach to static economics," *Journal of Political Economy*, **54** (1946), 97–115.

Kaysen, C., "A revolution in economic theory?," *Review of Economic Studies*, **14/1** (1946–47), 1–15.

Stone, R., "The theory of games," *Economic Journal*, **58** (1948), 185–201.

Copeland, A. H., Review of "The Theory of Games and Economic Behavior," *Bulletin of the American Mathematical Society*, **51** (1945), 498–504.

4. While this assumption seems natural enough for the primary interpretation in terms of parlor games, it is quite a subtle step in an economic model from

preferences to measurable utility or money. See, for example, the Appendix and Section 3 of the book of von Neumann and Morgenstern cited above.

The reader may find it profitable at this point to gain a comprehensive non-technical view of the theory of games from the following more or less popular articles:

McDonald, J., "Poker: an American game," *Fortune*, **37** (1948), 128–131 and 181–187.

McDonald, J., "The theory of strategy," *Fortune*, **38** (1949), 100–110.

McDonald, J., *Strategy in Poker, Business, and War*, W.W. Norton, New York, 1950.

Morgenstern, O., "The theory of games," *Scientific American* (May, 1949), 22–25.

Bellman, R., Blackwell, D., "Red dog, blackjack, and poker," *Scientific American* (January, 1951), 44–47.

Dresher, M., "Games of strategy," *Math. Magazine*, **25** (1951), 93–99.

Chapter Two

Matrix Games

2.1 Two Examples

The well-known game of Matching Pennies provides us with an example of a zero-sum two-person finite game with two moves. (Note that in this game, as in later examples, in order that only one player act at a move, we divide one natural simultaneous move into two consecutive moves.) In the first move, P_1 chooses the alternative "heads" or "tails," and in the second move, P_2, in ignorance of P_1's choice, chooses the alternative "heads" or "tails." After the choices have been made, P_2 pays P_1 the amount 1 if they match or -1 if they do not match. We can summarize the rules without losing any essential information by the diagram:

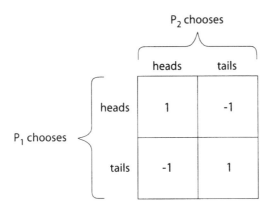

At the end of a play P_2 pays P_1 the amount shown. This diagram can be simplified considerably if we adopt the following conventions:

(a) The rows (columns) of the matrix will correspond to possible choices for P_1 (P_2).
(b) Since there are only two moves in the game under consideration, a choice for P_1 and a choice for P_2, that is, a row and a column, determine a play uniquely. The matrix entry at the corresponding intersection gives the amount that P_2 pays P_1 at the end of the play.

Hence we can replace Diagram 1 by

$$\begin{pmatrix} 1 & -1 \\ -1 & 1 \end{pmatrix}$$

and this matrix gives all of the essential information without any of the cumbersome names for the apparatus of the particular game.

In searching for how P_1 would play, it seems that there is no difference between the two modes of play open to him; in either case he will receive 1 or pay 1 depending on whether P_2 matches his choice or not. However, it is of the utmost importance that he conceal his choice from P_2, for, if P_2 can discover his choice in advance, P_2 can make the opposite choice and win the amount 1. One way for P_1 to make the prediction of his choice more difficult is to use a chance device. Suppose, for example, that P_1 makes his choice by flipping an unbiased coin, choosing heads if the coin shows heads and tails if the coin shows tails. By this means, the probability that P_1 will show heads is $1/2$ and the probability that he will show tails is $1/2$. Assuming that he makes his choice in this manner what can he expect to win on a single play? In case P_2 plays heads, the mathematical expectation [1] of P_1 is $(1/2)(1) + (1/2)(-1) = 0$, while if P_2 plays tails, it is $(1/2)(-1) + (1/2)(1) = 0$ and hence his expectation is 0. Indeed this is the only way that P_1 can play the game without leaving himself exposed to a negative expectation. To show this, suppose that he plays heads with probability x and tails with probability $1 - x$, where $0 \leqq x \leqq 1$. Then his expectation against heads played by P_2 is $x(1) + (1 - x)(-1) = 2x - 1$ and his expectation against tails played by P_2 is $x(-1) + (1 - x)(1) = 1 - 2x$. But, when $0 \leq x < 1/2$, $2x - 1 < 0$ and when $1/2 < x \leq 1$, $1 - 2x < 0$ and hence, if P_1 wishes to protect himself against a possible negative expectation, he must play heads and tails with equal probabilities. If he plays in this manner, his expectation is 0. It is clear and easily verified that the same conclusion holds for P_2.

A somewhat more subtle example is provided by the following matching game.

A SKIN GAME [2]. The two players are each provided with an ace of diamonds and an ace of clubs. Player P_1 is given the two of diamonds while P_2 is given the two of clubs. In the first move, P_1 shows one of his cards and in the second move, P_2, in ignorance of P_1's choice, shows one of his. Player P_1 wins if the *suits* match, P_2 if they do not. The amount of the payoff is the numerical value of the card shown by the winner. If the two deuces are shown the payoff is zero.

With our previous conventions, (a) and (b), this game can be described by the following matrix:

$$\begin{pmatrix} 1 & -1 & -2 \\ -1 & 1 & 1 \\ 2 & -1 & 0 \end{pmatrix}$$

On the surface it appears to be a fair game, which is to say that the expectation of the players for optimal play appear to be zero, by the apparent symmetry of the possibilities.

However, let us assume that we are P_1 hunting for an optimal way to play. Looking at his possible modes of play, it appears that he might as well choose the third row as choose the first row, since if P_2 counters with the second column he loses the same amount, while if P_2 counters with the first or third column he actually fares better. So let us assume that P_1 never uses the first row, chooses the second with probability x, and, consequently, chooses the third with probability $1 - x$. Then we can plot his expectation against the three possible modes of play for P_2 as a function of x.

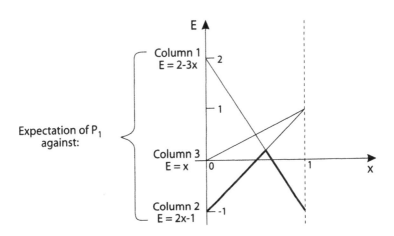

Now, being a pessimistic or perhaps prudent player, P_1 realizes that, for each choice of a probability x, he can only count on the expectation which is the minimum of the three possible results. This is shown in our graph by the heavy line. Adopting this point of view, he will choose the x which maximizes this minimum. It is clear from the graph that this maximum occurs at the point where $2 - 3x = 2x - 1$, that is, where $x = 3/5$, and at this point P_1's expectation is $1/5$ if P_2 chooses his first or second modes of play while it is $3/5$ if P_2 is foolish enough to use his third mode.

On the other hand, let us examine how well P_2 can do. From the above discussion it seems reasonable to assume that he will not use his third column and that he chooses the first and second columns with probabilities y and $1 - y$ respectively. Again we plot his expectation against the three possible modes of play for P_1.

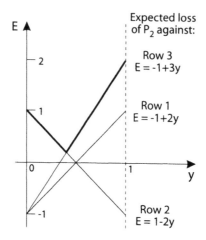

Since our expectation always measures the amount that P_2 pays P_1 at the end of a play, to be safe P_2 must assume he loses the maximal expectation for each y; again this is shown by a heavy line on the graph. Since his aim is to minimize his expected loss, he chooses the point where $1 - 2y = 3y - 1$, that is, where $y = 2/5$. At this point, his expected loss is $1/5$ if P_1 chooses his second or third modes of play, while he can expect to win $1/5$ if P_1 is imprudent and plays his first mode.

Summarizing, we have found probabilities for P_1 which assure him an expected gain of at least $1/5$ and probabilities for P_2 which protect him from an expected loss of more than $1/5$. It is clear that these ways of play have a strong claim on the title of giving the "most favorable result" to the two players.

EXERCISE 1. Analyze Modified Matching Pennies defined by the matrix

$$\begin{pmatrix} 2 & -1 \\ -1 & 1 \end{pmatrix}$$

where an extra unit is won by P_1 if heads are matched.

EXERCISE 2. Find the best way to play the game defined by

$$\begin{pmatrix} 0 & c & -b \\ -c & 0 & a \\ b & -a & 0 \end{pmatrix}$$

where a, b, c are positive numbers. (Hint: Note the similarity to the game of Stone, Paper, Scissors where $a = b = c$.)

(The exercises above, as those to be found later in these notes, should be worked in order and with the methods available at the point at which they occur.)

2.2 The Definition of a Matrix Game

With the two examples of Section 1 in mind, we can now define matrix games and their solutions. We shall first state a formal definition and then give it its interpretation.

DEFINITION I. A *matrix game* Γ is given by any $m \times n$ matrix

$$A = \begin{pmatrix} a_{11} & a_{12} & \cdots & a_{1n} \\ a_{21} & a_{22} & \cdots & a_{2n} \\ \cdots & \cdots & \cdots & \cdots \\ a_{m1} & a_{m2} & \cdots & a_{mn} \end{pmatrix}$$

in which the entries a_{ij} are real numbers. By a *mixed strategy for* P_1 we shall mean an ordered m-tuple $X = (x_1, \ldots, x_m)$ of non-negative real numbers x_i such that $x_1 + \cdots + x_m = 1$. Similarly, a *mixed strategy for* P_2 will mean an ordered n-tuple $Y = (y_1, \ldots, y_n)$ of non-negative real numbers y_j such that $y_1 + \cdots + y_n = 1$. For each $i = 1, \ldots, m$, the mixed strategy which is 1 in the ith component and 0 elsewhere is called the ith *pure strategy for* P_1 and will be denoted by i when no confusion results. Similarly j will denote an n-tuple which is 1 in the jth component and 0 elsewhere and is called the jth *pure strategy for* P_2. The *payoff function* for Γ is defined to be

$$E(X, Y) = \sum_{i,j} x_i a_{ij} y_j$$

where $X = (x_1, \ldots, x_m)$ and $Y = (y_1, \ldots, y_n)$ are mixed strategies. A *solution* of Γ is a pair of mixed strategies $\bar{X} = (\bar{x}_1, \ldots, \bar{x}_m)$, $\bar{Y} = (\bar{y}_1, \ldots, \bar{y}_n)$, and a real number v such that:

$$E(\bar{X}, j) \geqq v$$

for the pure strategies $j = 1, \ldots, n$

$$E\left(i, \bar{Y}\right) \leqq v$$

for the pure strategies $i = 1, \ldots, m$. The $\bar{X}$ and $\bar{Y}$ are called *optimal strategies* and the number v is called the *value* of the game.

The interpretation of this definition is clear after the two preceding examples. The rows (columns) of A correspond to alternatives for P_1 (P_2) which are now called pure strategies. The game is a two-move game in which P_2 makes his choice in ignorance of the choice of P_1. If P_1 has chosen his i^{th} pure strategy and P_2 his j^{th} pure strategy then P_2 pays P_1 the amount a_{ij}. If P_1 and P_2 play their pure strategies with the probabilities indicated by the mixed strategies X and Y then P_2 can expect to pay the amount $E(X, Y)$ to P_1.

The two keystones of the notion of solution are expectation and safety. The amount v is the expectation that P_1 can assure himself no matter what P_2 does while P_2 can protect himself against larger expectations than v. The remarkable fact which provided the starting point for all investigations of matrix games was proved by von Neumann as early as 1928 [3]. It is the

FUNDAMENTAL THEOREM. Every matrix game has a solution.

2.3 The Fundamental Theorem for 2 × 2 Matrix Games

As motivation for the general case we will prove the fundamental theorem for 2×2 games in this section. In order to soften the abrupt transition to new concepts and vector notation this theorem will be rephrased as an algebraic theorem without regard to the game terminology. The reader should make the necessary connections.

THEOREM I. Given the 2×2 matrix

$$a = \begin{pmatrix} a_{11} & a_{12} \\ a_{21} & a_{22} \end{pmatrix}$$

there exist vectors $\bar{X} = (\bar{x}_1, \bar{x}_2)$, where $\bar{x}_1 \geq 0, \bar{x}_2 \geq 0$ and $\bar{x}_1 + \bar{x}_2 = 1$, and $\bar{y} = (\bar{y}_1, \bar{y}_2)$, where $\bar{y}_1 \geq 0, \bar{y}_2 \geq 0$, and $\bar{y}_1 + \bar{y}_2 = 1$, and a real number v such that:

(1) $\bar{x}_1 a_{11} + \bar{x}_2 a_{21} \geq v$

$\bar{x}_1 a_{12} + \bar{x}_2 a_{22} \geq v$

(2) $a_{11} \bar{y}_1 + a_{12} \bar{y}_2 \leq v$

$a_{21} \bar{y}_1 + a_{22} \bar{y}_2 \leq v$

PROOF. If P_2 plays the mixed strategy $Y = (y_1, y_2)$ he can expect to pay $E(1, Y) = a_{11}y_1 + a_{12}y_2$ against 1 by P_1 and $E(2, Y) = a_{21}y_1 + a_{22}y_2$ against 2 by P_1. Let us now plot these expectations as the coordinates of a point in a 2-dimensional space which we may call P_2's "expectation space." Since Y satisfies the conditions $y_1 \geq 0, y_2 \geq 0$, and $y_1 + y_2 = 1$, the set of points obtained from all possible mixed strategies is the line segment joining the point (a_{11}, a_{21}) to the point (a_{12}, a_{22}).

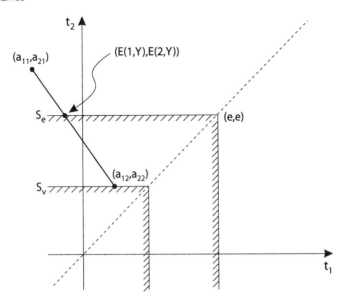

Game-wise, P_2 is assured of paying no more than the expectation e if and only if his expectation point lies in the set S_e of all points with both coordinates less than or equal to e. It is clear that we stand the best chance of satisfying (2) or, alternatively, P_2 is playing optimally, if we choose a Y with the smallest possible e. Geometrically, this means translating S_e until it is just in contact with the segment. Call this corner S_v and let $\bar{Y}$ be any Y which yields a point on the segment in the contact. Clearly (2) is satisfied for this choice of $\bar{Y}$ and v.

Now we want a mixed strategy $\bar{X} = (\bar{x}_1, \bar{x}_2)$ satisfying (1) for this v. Assume that we have found such an $\bar{X}$ and consider the line $\bar{x}_1 t_1 + \bar{x}_2 t_2 = v$. On the one hand we have $\bar{x}_1 t_1 + \bar{x}_2 t_2 \geq v$ for all points (t_1, t_2) on the segment by (1) and on the other hand we have $\bar{x}_1 t_1 + \bar{x}_2 t_2 \leq \bar{x}_1 v + \bar{x}_2 v = v$ for all (t_1, t_2) in S_v by its definition. We shall express this by saying that the line *separates* the segment from the set S_v. (In general, a line $at_1 + bt_2 = c$ *separates* the sets A and B if $at_1 + bt_2 \geq c$ for all (t_1, t_2) in A and $at_1 + bt_2 \leq c$ for all (t_1, t_2) in B.) Thus we find that any optimal $\bar{X}$, if one exists, defines a separating line and we suspect that, conversely, any separating line will define an optimal $\bar{X}$.

Let us assume that $at_1 + bt_2 = c$ separates the segment from S_v, that is,

(3) $aa_{11} + ba_{21} \geq c$

(4) $aa_{12} + ba_{22} \geq c$

(5) $at_1 + bt_2 \leq c$ for all (t_1, t_2) in S_v.

Now, since $(v - 1, v)$ and $(v, v - 1)$ are in S_v, we have

(6) $a(v - 1) + bv \leq c$

(7) $av + b(v - 1) \leqq c$

while, since it is clear geometrically that the point (v, v) lies on any such line,

(8) $av + bv = c$.

Subtracting (8) from (6) and (7) we have

(9) $a \geqq 0$ and $b \geqq 0$.

However, not both a and b can be zero and so we can define $\bar{x}_1 = a/(a + b)$, $\bar{x}_2 = b/(a + b)$ so that $\bar{x}_1 \geq 0$, $\bar{x}_2 \geq 0$ and $\bar{x}_1 + \bar{x}_2 = 1$. But then (3) and (4) combined with (8) yield the desired inequalities (1).

Thus, summing up: to find an optimal $\bar{Y}$, choose the corner S_v which is in contact with the segment. Then any $\bar{X}$ with non-negative components summing to 1 which is the normal to a line which separates the segment from S_v is an optimal strategy for P_1; any $\bar{Y}$ that yields a point of the segment which is in the contact is an optimal strategy for P_2; and v is the value of the game.

Here we have the bare ideas for a proof of the fundamental theorem for any matrix game. In general, the segment will be replaced by a polyhedral convex set C and the proof will be based on the fact that C can always be separated from the "corner" by a hyperplane. This and other facts about convex sets will be considered in the next section.

2.4 The Geometry of Convex Sets

In this section we will investigate some of the simpler properties of convex sets in n-dimensional Euclidean space. Our choice of topics will be motivated by the contemplated application to the theory of games. Therefore our treatment will by no means be a complete survey of the field of convex sets [4].

The representation of points in the Cartesian plane by ordered pairs of numbers $T = (t_1, t_2)$ is familiar to all students of analytic geometry. The objects T can be regarded geometrically in two ways: (1) as a *point* with coordinates t_1, t_2 in a fixed coordinate system, and (2) as a *vector* or translation by the amounts t_1, t_2 in two fixed coordinate directions. These two attitudes are related by the fact that the vector T corresponds to the translation of the origin to the point T. Although we shall usually speak of vectors, this relation allows us to use the word point instead when it seems more natural.

With an eye to generalizing certain elementary properties of the 2-dimensional Euclidean space R_2, let us catalogue those properties of vectors in R_2 that appear to be important. The reader should notice that all of these properties can be stated without reference to the coordinate system and we gain considerably in economy of expression by *not* referring to the coordinate system.

(A) ADDITION OF VECTORS. To every pair T and U of vectors in R_2 there corresponds a vector V, called the sum of T and U, written $V = T + U$, and such that:

1. addition is *commutative*, i.e., $T + U = U + T$;
2. addition is *associative*, i.e., $(T + U) + V = T + (U + V)$;
3. there exists in R_2 a unique vector 0, called the *origin*, such that $T + 0 = T$ for all T in R_2;
4. to each T in R_2 there corresponds a unique vector, called the *inverse* of T and denoted by $-T$, such that $T + (-T) = 0$.

(B) MULTIPLICATION OF VECTORS BY REAL NUMBERS. To every pair, a and T, where a is a real number and T is a vector, there corresponds a vector U, called the *product* of a and T, written $U = aT$, and such that:

1. multiplication is *distributive* with respect to the addition of vectors, i.e., $a(T + U) = aT + aU$;
2. multiplication is *distributive* with respect to the addition of numbers, i.e., $(a + b)T = aT + bT$;
3. multiplication is *associative* with respect to multiplication of numbers, i.e., $a(bT) = (ab)T$;
4. $0T = 0$ and $1T = T$. Note that the symbols appearing in the first equation are conceptually different; the symbol on the left is the number zero while the symbol on the right is the origin of R_2.

(C) LENGTH. To every vector T in R_2 there corresponds a real number, called the length of T, written $\|T\|$, and such that:

1. $\|T\| \geqq 0$ and $\|T\| = 0$ if and only if $T = 0$;
2. $\|aT\| = |a| \|T\|$;
3. the length satisfies the "triangle inequality," i.e., $\|T + U\| \leqq \|T\| + \|U\|$ for every pair T and U of vectors in R_2.

We are also able to talk about the angles between vectors in R_2; however, in practice, it is more convenient to use the cosine of the angle between two vectors, reflecting the fact that the angle is the length of a circular segment, while the cosine is the length of a line segment. Moreover, we can introduce one concept which subsumes both angle and length.

(D) INNER PRODUCT OF VECTORS. To every pair T and U of vectors in R_2 there corresponds a real number, called the inner product of T and U, written $T \cdot U$, and such that:

1. the inner product is *commutative*, i.e., $T \cdot U = U \cdot T$;

2. the inner product is *bilinear*, i.e., $(aT + bU) \cdot V = a(T \cdot V) + b(U \cdot V)$ for all pairs of numbers a and b and all triples of vectors T, U, and V in R_2;

3. $T \cdot T \geq 0$ and $T \cdot T = 0$ if and only if $T = 0$.

EXERCISE 3. Prove properties $C1$, $C2$, $C3$ from $D1$, $D2$, $D3$ and the definition $\| T \| = +\sqrt{T \cdot T}$.

Finally, we must characterize the dimension of R_2.

DEFINITION 2. A set of vectors $T_1, \ldots, T_n$ is called *linearly independent* if, for all numbers $a_1, \ldots, a_n$, $a_1 T_1 + a_2 T_2 + \cdots + a_n T_n = 0$ implies $a_1 = a_2 = \cdots = a_n = 0$. Otherwise it is called *linearly dependent*.

Then the property of R_2 which makes it 2-dimensional is (E_2): There exist 2 linearly independent vectors while every set of 3 vectors is linearly dependent. One notices immediately that, with the exception of E_2, nothing in our catalogue refers to the dimension of R_2, and we are in a position to generalize to any number of dimensions.

DEFINITION 3. An *n-dimensional Euclidean space R_n* is a set of objects called vectors which satisfy A, B, D, and E_n: there exist n linearly independent vectors while every set of $n + 1$ vectors is linearly dependent.

It can be shown that there is essentially only one system satisfying these requirements and that it can be realized as ordered *n*-tuples $T = (t_1, \ldots, t_n)$ of real numbers with the following definitions of the operations of the axioms:

$$(t_1, \ldots, t_n) + (u_1, \ldots, u_n) = (t_1 + u_1, \ldots, t_n + u_n)$$

$$a(t_1, \ldots, t_n) = (at_1, \ldots, at_n)$$

$$(t_1, \ldots, t_n) \cdot (u_1, \ldots, u_n) = t_1 u_1 + \ldots + t_n u_n .$$

EXERCISE 4. Prove A, B, C, and E_2 for ordered number couples (t_1, t_2) using the definitions of the operations given above.

One linear condition in R_2 (such as $x_1 t_1 + x_2 t_2 = a$ where x_1, x_2 and a are constant) defines a line; one linear condition in R_3 defines a plane; the corresponding object defined by one linear condition in R_n is called a hyperplane.

DEFINITION 4. A *hyperplane $H(X, a)$* in R_n is the set of all vectors T such that $X \cdot T = a$ for a given vector $X \neq 0$ and real number a.

It can be shown that H is an $(n - 1)$-dimensional Euclidean space (although there is no natural way to choose an origin unless $a = 0$ and hence 0 is in H) and it clearly divides R_n into two half-spaces which we will denote by

$$H^+(X, a) \;=\; \{T \,|\, X \cdot T \geqq a\}$$

$$H^-(X, a) \;=\; \{T \,|\, X \cdot T \leqq a\}.$$

Among the sets of R_n which will interest us particularly are the convex sets. The concept of a convex set undoubtedly originated in the solution of the physical problem of finding the center of gravity of m given points $T_1, \ldots, T_m$ at which we have masses $a_1, \ldots, a_m$. The solution of this problem is well known; if we denote the center of gravity by T then

$$T \;=\; \frac{a_1 T_1 + \cdots + a_m T_m}{a_1 + \cdots + a_m}$$

or, letting $b_i = \dfrac{a_i}{a_1 + \cdots + a_m}$ for $i = 1, \ldots, m$, we have $T = b_1 T_1 + \cdots + b_m T_m$ where all $b_i \geqq 0$ and $\sum_i b_i = 1$.

DEFINITION 5. A subset C of R_n is called *convex* if, whenever $T_1, \ldots, T_m$ are in C and $b_1, \ldots, b_m$ are non-negative numbers that sum to one, $T = b_1 T_1 + \cdots + b_m T_m$ is in C. The vector T is called a *convex combination* of $T_1, \ldots, T_m$.

Clearly, we can replace this condition by the single condition for $m = 2$. For, assuming that we have shown all convex combinations of $m - 1$ points in C are in C, suppose $T = b_1 T_1 + \cdots + b_m T_m$ is a convex combination of points $T_1, \ldots, T_m$ from C. Then, since $b_1 + \cdots + b_m = 1$, we can assume that $b_1 \neq 0$ and set

$$b_i' \;=\; \frac{b_i}{b_1 + \cdots + b_{m-1}} \qquad \text{for } i = 1, \ldots, m - 1.$$

Hence $T = (b_1 + \cdots + b_{m-1})(b_1' T_1 + \cdots + b_{m-1}' T_{m-1}) + b_m T_m$, which exhibits T as a convex combination of two points in C and proves that T is in C.

CRITERION. A set C is convex if and only if, for all pairs of vectors T and U in C and all pairs a and b of non-negative numbers such that $a + b = 1$, $aT + bU$ is in C.

Geometrically, this says that a convex set is one which contains all line segments joining two points in the set.

DEFINITION 6. The *convex hull* $C(S)$ of a given set S is the set of all convex combinations of sets of points from S.

Remark that the convex hull of S is simply the smallest convex set containing S (since the intersection of any number of convex sets is convex and hence it is meaningful to talk about the smallest convex set containing S).

The next theorem which we will prove may well be called the keystone of the geometry of convex sets; it has a long history in the mathematical literature [5] and has been given the name "the theorem of the supporting hyperplane".

THEOREM 2. Given a closed [6] convex set C in R_n and a vector U not in C, there exists a hyperplane H such that:

1. C is contained in H^+;
2. U is in H^- but not in H.

PROOF. Choose the point V of C which is closest to U. (The existence of such a point depends upon the fact that a continuous function on a closed bounded set in R_n assumes a minimum value.) The continuous function in question is $\|T - U\|$; it is defined on the closed bounded set consisting of those T in C for which $\|T - U\| \leq \|T_0 - U\|$ for some fixed T_0 in C.) To simplify the notation assume that $V = 0$.

Consider the hyperplane $H(-U, 0)$ which consists of those T for which $-U \cdot T = 0$. It is clear that condition (2) of the theorem is met since $-U \cdot U < 0$ by Axiom $D3$ for the space R_n. I contend that (1) is met also, i.e., $C \subset H^+$. If this is not so, there must be a vector T in C for which $-U \cdot T < 0$. i.e., for which $U \cdot T > 0$. We dispose of this possibility in two cases.

CASE 1.

$$\frac{U \cdot T}{T \cdot T} \geq 1 \implies U \cdot T \geq T \cdot T \implies \|T - U\|^2 = T \cdot T - 2U \cdot T + U \cdot U$$
$$= U \cdot U - (U \cdot T - T \cdot T) - U \cdot T < U \cdot U = \|U\|^2$$
$$\implies \|T - U\| < \|U\| = \|0 - U\|,$$

contradicting the fact that 0 is the point of C closest to U.

CASE 2.

$$0 < \frac{U \cdot T}{T \cdot T} < 1 \implies \left(\frac{U \cdot T}{T \cdot T}\right) T \text{ is in } C$$

and

$$\left\|\left(\frac{U \cdot T}{T \cdot T}\right) T - U\right\|^2 = U \cdot U - \left(\frac{U \cdot T}{T \cdot T}\right)(U \cdot T) < \|U\|^2$$
$$\implies \left\|\left(\frac{U \cdot T}{T \cdot T}\right) T - U\right\| < \|U\|,$$

that is, $((U \cdot T)/(T \cdot T))T$ is in C and is closer to U than 0, which is a contradiction. This completes the proof. The following figure explains the geometric line of thought behind the algebraic proof that has been given.

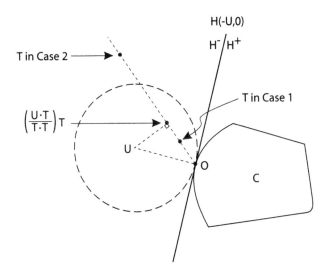

DEFINITION 7. If H is a hyperplane and C is a convex set such that C is contained in H^+, then H is called a *supporting hyperplane* for C and H^+ is called a *support* for C.

COROLLARY 1. Any closed convex set is the intersection of its supports.

PROOF. Let $\bigcap H^+$ denote the intersection of all supports for C. Clearly, $\bigcap H^+$ contains C. However, if U is not in C then Theorem 2 asserts that there is a support H^+ for C which does not contain U. Therefore, $\bigcap H^+$ is contained in C and hence $C = \bigcap H^+$. Remark that if C has no supports then we must set $\bigcap H^+ = R_n$.

COROLLARY 2. Given a closed convex set C such that $C \neq R_n$, there exists a support for C which contains a vector in C.

PROOF. Immediate consequence of Theorem 2.

To extend Theorem 2 to an arbitrary convex set in R_n we need a theorem which expresses a property peculiar to convex sets in a finite dimensional space [7].

DEFINITION 8. The point U is an *interior point* of a set S in R_n if there exists an $\epsilon > 0$ such that $\parallel V - U \parallel < \epsilon$ implies V is in S.

EXERCISE 5. Let the set S in R_n consist of the origin 0 and n linearly independent vectors $T_1, \ldots, T_n$. Show that $C(S)$, the convex hull of S, contains an interior point. (Hint: Sharpen Corollary 1 to show that $C(S)$ is the intersection of its *extreme supports*, i.e., those which contain n points from the set S.)

THEOREM 3. Every convex set C in R_n either contains an interior point or is contained in a hyperplane.

PROOF. For notational simplicity assume that 0 is in C. Then choose a system of linearly independent vectors $T_1, \ldots, T_d$ from C such that every set of $d + 1$ vectors from C is linearly dependent. Then there are two relevant cases.

CASE 1. If $d = n$, then Exercise 5 says that C contains an interior point.

CASE 2. If $d < n$, then the subspace R_d generated by $T_1, \ldots, T_d$ is not all of R_n and we can choose a non-zero vector X such that $X \cdot T_k = 0$ for $k = 1, \ldots, d$ [9]. But, if T is any vector in C, we must have $T = a_1 T_1 + \cdots + a_d T_d$ since the set $T_1, \ldots, T_d, T$ is linearly dependent while $T_1, \ldots, T_d$ is not. Taking inner products with X, $X \cdot T = 0$ gives $C \subset H(X, 0)$ and the theorem is proved.

THEOREM 4. Given *any* convex set C with $C \neq R_n$ there exists a support for C.

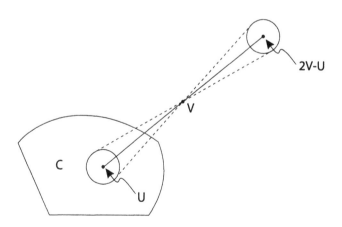

PROOF. First we enlarge C to D, the smallest closed set containing C, by adding to C all of its limit points. It is obvious that D is convex for, if U and V are limit points of C then we have two sequences $U_1, U_2, \ldots$ and $V_1, V_2, \ldots$ of vectors in C which tend to U and V respectively. Then, to show that every convex combination

$aU + bV$ of U and V is a limit point of C, we consider the sequence $aU_1 + bV_1$, $aU_2 + bV_2, \ldots$. We have

$$\begin{aligned}
\|(aU + bV) - (aU_k + bV_k)\| &= \|a(U - U_k) + b(V - V_k)\| \\
&\leqq a\|U - U_k\| + b\|V - V_k\|
\end{aligned}$$

and hence, as $U_k \longrightarrow U$ and $V_k \longrightarrow V$, $aU_k + bV_k \longrightarrow aU + bV$.

Now, by Theorem 2, it clearly suffices to show $D \neq R_n$. But consider the alternatives presented by Theorem 3. If C is contained in a hyperplane H then all of its limit points are in H and hence D is contained in H. If C has an interior point U and there is a point V not in C then for some $\epsilon > 0$ all the points within ϵ of $2V - U$ are not in C and hence $2V - U$ is not in D (see figure).

Now consider the problem of separating two arbitrary convex sets C and D. Recall:

DEFINITION 9. The hyperplane $H(X, a)$ is said to *separate* the set C from the set D if $C \subset H^+$ and $D \subset H^-$.

In the following theorem we will assume that we have two convex sets C and D which intersect at the origin and such that D has an interior point but no point of C is interior to D. The latter restrictions are made to eliminate the two types of situation in which no separation is possible.

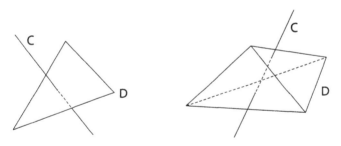

THEOREM 5. Let two convex sets C and D be such that:

(1) 0 lies in both C and D;
(2) D has an interior point W;
(3) no point of C is interior to D.

Then there exists a hyperplane $H(X, 0)$ which separates C from D.

PROOF. Let E be the set of all $aT - bU$, where a and b are any non-negative numbers, T is any vector in C, and U is any vector in D. The proof can be given in several steps.

(a) It is clear that E is a convex set.

(b) Any supporting hyperplane for E which passes through the origin, $H(X, 0)$, separates C from D. For, taking $a = 1$ and $b = 0$, $X \cdot T \geq 0$ for all T in C and, taking $a = 0$ and $b = 1$, $X \cdot U \leq 0$ for all U in D.

(c) The point W is not in E. For, suppose $W = aT - bU$ where $a \geq 0, b \geq 0$, T is in C, and U is in D. Then

$$\frac{1}{1+b} W + \frac{b}{1+b} U = \frac{a}{1+b} T.$$

If $a/(1 + b) \leq 1$ then $(a/(1 + b))T$ is a convex combination of 0 and T, hence lies in C, and is a convex combination of U in D and W which is interior to D, hence is interior to D.

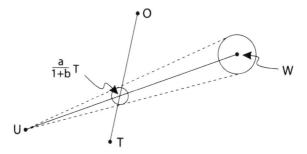

On the other hand, if $a/(1 + b) \geq 1$, then T is in C and is a convex combination of 0 and $(a/(1 + b))T$, which is interior to D, hence is interior to D.

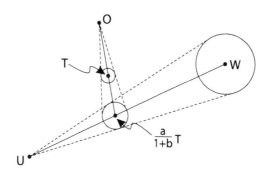

Now, by (a), (c), and Theorem 4, there is a supporting hyperplane $H(X, a)$ for E and I contend that $H(X, 0)$ also supports E. For suppose $X \cdot V < 0$ and $X \cdot V \geq a$ for some V in E. Then, for some $d > 0$, $d(X \cdot V) + X \cdot V < a$. Hence, $X \cdot (dV + V) < a$ and $dV + V = (d + 1)V$ is clearly in E which is a contradiction.

Therefore, by (b), $H(X, 0)$ separates C from D.

2.5 Fundamental Theorem for All Matrix Games

We are now in a position to prove the fundamental theorem for all matrix games. Our proof parallels that given for 2×2 matrices.

THEOREM 6. Given any $m \times n$ matrix

$$A = (a_{ij}), \text{ there exist vectors}$$

$$\bar{X} = (\bar{x}_1, \ldots, \bar{x}_m), \quad \text{all } \bar{x}_1 \geq 0, \ \bar{x}_1 + \cdots + \bar{x}_m = 1,$$

$$\bar{Y} = (\bar{y}_1, \ldots, \bar{y}_n), \quad \text{all } \bar{y}_j \geq 0, \ \bar{y}_1 + \cdots + \bar{y}_n = 1,$$

and a real number v such that

(1) $\bar{x}_1 a_{1j} + \cdots + \bar{x}_m a_{mj} \geq v$ for $j = 1, \ldots, n$,

(2) $a_{i1} \bar{y}_1 + \cdots + a_{in} \bar{y}_n \leq v$ for $i = 1, \ldots, m$.

PROOF. We work in P_2's expectation space R_m and consider the set C of all vectors $T = y_1 T_1 + \cdots + y_n T_n$ where $T_j = (a_{1j}, \ldots, a_{mj})$, all $y_j \geq 0$, and $y_1 + \cdots + y_n = 1$. The set C is closed and convex, indeed, is the convex hull of the points $T_1, \ldots, T_n$. If $G = (t_1, \ldots, t_m)$ is a point consider the function $E(T) = \max_{i=1,\ldots,m} \{t_i\}$. This is a continuous function defined on the closed and bounded set C in R_m and hence assumes its minimum at a point $\bar{T} = \bar{y}_1 T_1 + \cdots + \bar{y}_n T_n$ in C. Let $v = E(\bar{T})$; it is clear that (2) is satisfied for this choice of $\bar{Y} = (\bar{y}_1, \ldots, \bar{y}_n)$ and v.

Now define D to be the set of all $T = (t_1, \ldots, t_m')$ such that $t_i \leq v$ for $i = 1, \ldots, m$. The set D intersects C at $\bar{T}$ and has interior points, but no point of C is interior to D. (Otherwise there would be a T in C with all $t_i < v$, contradicting the choice of v.) Hence, by Theorem 5, there is a hyperplane $H(X, a)$ which separates C from D.

I contend that the point $V = (v, \ldots, v)$ lies in H. For if $X \cdot V < a$ then

$$X \cdot (2\bar{T} - V) = 2X \cdot T - X \cdot V < 2a - a = a$$

which contradicts the fact that $2\bar{T} - V$ is in D. Hence $X \cdot V = a$ and V is in H.

Let V_i be the vector which has component v except at the ith coordinate where it is $v - 1$, for $i = 1, \ldots, m$. Then V_i is in D and hence $X \cdot V_i = (X \cdot V) - x_i \leq a$. Therefore, $x_i \geq 0$ for $i = 1, \ldots, m$. Since not all the x_i can be 0 we can define $\bar{X} = (\bar{x}_1, \ldots, \bar{x}_m)$ where

$$\bar{x}_i = \frac{x_i}{x_1 + \cdots + x_m} \quad \text{for } i = 1, \ldots, m.$$

Then all $\bar{x}_i \geq 0$, $\bar{x}_1 + \cdots + \bar{x}_m = 1$ and, since C is contained in H^+,

$$\bar{X} \cdot T_j = \frac{1}{x_1 + \cdots + x_m} X \cdot T_j \geq \frac{a}{x_1 + \cdots + x_m} = v$$

for $j = 1, \ldots, n$, which is condition (1) of the theorem.

Thus, summing up: to find an optimal $\bar{Y}$, choose the corner S_v which is in contact with the convex set C. Then any $\bar{Y}$ which corresponds to a point of C in the contact is an optimal strategy for P_2; any $\bar{X}$ with unit component sum which is normal to a hyperplane separating C from S_v is an optimal strategy for P_1; the number v for which contact takes place is the value of the game.

Theorem 6 occupies such an important position in the theory of zero-sum two-person finite games that it behooves us to become familiar with alternative statements.

SADDLEPOINT STATEMENT. For all matrix games, there exist mixed strategies $\bar{X}$ and $\bar{Y}$ such that

$$E(X, \bar{Y}) \leqq E(\bar{X}, \bar{Y}) \leqq E(\bar{X}, Y)$$

for all mixed strategies X, Y.

MINIMAX STATEMENT. For all matrix games the quantities

$$\min_Y \max_X E(X, Y) \text{ and } \max_X \min_Y E(X, Y)$$

exist and are equal.

PROOF. We shall prove Theorem 6 $\Longrightarrow$ Saddlepoint Statement $\Longrightarrow$ Minimax Statement $\Longrightarrow$ Theorem 6.

Theorem 6 $\Longrightarrow$ Saddlepoint Statement. Let $\bar{X}$ and $\bar{Y}$ be optimal strategies. Then

$$E(i, \bar{Y}) \leqq v \quad \text{for } i = 1, \dots, m \,.$$

Hence, for all $X = (x_1, \dots, x_m)$ with all $x_i \geqq 0$, $x_1 + \cdots + x_m = 1$,

$$E(X, \bar{Y}) \leqq x_1 v + \cdots + x_m v = v.$$

Similarly,

$$E(\bar{X}, Y) \geqq v$$

for all mixed strategies Y. With $X = \bar{X}$ in the first inequality and $Y = \bar{Y}$ in the second, $E(\bar{X}, \bar{Y}) = v$, and

$$E(X, \bar{Y}) \leqq E(\bar{X}, \bar{Y}) \leqq E(\bar{X}, Y)$$

for all mixed strategies X and Y.

Saddlepoint Statement $\Longrightarrow$ Minimax Statement. By the Saddlepoint Statement, $E(X, \bar{Y}) \leq E(\bar{X}, Y)$ for all X and Y. Also, for a fixed choice of $\bar{Y}$, $E(X, \bar{Y})$ is a continuous function of X defined on the closed bounded set of all mixed strategies X in R_m and hence assumes a maximum. Clearly,

$$\max_X E(X, \bar{Y}) \leqq E(\bar{X}, Y).$$

But $E(\bar{X}, Y)$ is also a continuous function of Y defined on the closed bounded set of all mixed strategies Y in R_n and hence assumes a minimum. Obviously,

$$\max_{X} E(X, \bar{Y}) \leq \min_{Y} E(\bar{X}, Y).$$

Since $\bar{X}$ and $\bar{Y}$ are fixed choices of mixed strategies,

$$\min_{Y} \max_{X} E(X, Y) \leq \max_{X} E(X, \bar{Y}) \leq \min_{Y} E(\bar{X}, Y) \leq \max_{X} \min_{Y} E(X, Y).$$

On the other hand,

$$\max_{X} E(X, Y) \geq \min_{Y} E(X, Y)$$

for all choices of the Y which is left free on the left-hand side and of the X which is left free on the right-hand side. Hence

$$\min_{Y} \max_{X} E(X, Y) \geq \max_{X} \min_{Y} E(X, Y).$$

Combining the two lines of argument, we have

$$\min_{Y} \max_{X} E(X, Y) = \max_{X} \min_{Y} E(X, Y).$$

Minimax Statement $\Longrightarrow$ Theorem 6. Let v be the common value of $\min_Y \max_X E(X, Y)$ and $\max_X \min_Y E(X, Y)$ and choose $\bar{X}$ and $\bar{Y}$ so that

$$\max_{X} E(X, \bar{Y}) = v = \min_{Y} E(\bar{X}, Y).$$

Applying this equality to the particular mixed strategies X and Y which are pure strategies i and j, we have:

$$E(\bar{X}, j) \geq v \quad \text{for } j = 1, \ldots, n$$
$$E(i, \bar{Y}) \leq v \quad \text{for } i = 1, \ldots, m.$$

It is often useful to be able to change all of the payoffs by a fixed amount to adjust the value of the game. To this end we prove:

PROPOSITION 1. Let $A = (a_{ij})$, $B = (b_{ij})$ where $b_{ij} = a_{ij} + a$ for $i = 1, \ldots, m$ and $j = 1, \ldots, n$. Then, if we denote by v_A and v_B the values of the matrix games defined by A and B, $v_B = v_A + a$ and the games have the same sets of optimal strategies.

PROOF. If we note that $E_B(X, Y) = E_A(X, Y) + a$ by direct computation, the other assertions are easily verified.

Another proposition which finds constant use in the search for solutions says, in effect, that it is unwise to play a pure strategy that yields less than the value of the game against an optimal strategy for your opponent. Stated mathematically, this is:

PROPOSITION 2. Let $\bar{X}, \bar{Y}$, and v be a solution for the matrix game A. Then $E(\bar{X}, j) > v$ implies $\bar{y}_j = 0$ and $E(i, \bar{Y}) < v$ implies $\bar{x}_i = 0$.

PROOF. We prove the first statement only. If $E(\bar{X}, j) > v$ and $\bar{y}_j > 0$ then $\bar{y}_j E(\bar{X}, j) > \bar{y}_j v$. This would imply $E(\bar{X}, \bar{Y}) = \sum_j \bar{y}_j E(\bar{X}, j) > \bar{y}_1 v + \cdots + \bar{y}_n v = v$ which is a contradiction.

2.6 A Graphical Method of Solution

The geometrical characterization of the solutions of a matrix game given at the end of the proof of Theorem 6 provides the basis for solving all matrix games. Since we can certainly visualize the sets involved in the cases where one of the players has but 2 or 3 pure strategies, it may be well to strengthen our intuitions by considering the application of the method to several $3 \times n$ matrices (the $2 \times n$ case being too simple to illustrate any complexity).

If we consider a $3 \times n$ matrix game, the convex set C consists of all convex combinations of the n columns of the matrix considered as points in a 3-dimensional space with coordinates t_1, t_2, t_3. On the other hand, the "corner" S_v consists of all points $T = (t_1, t_2, t_3)$ such that $t_1 \leq v, t_2 \leq v, t_3 \leq v$ and is the translate of the "negative octant" S_0 along the line $t_1 = t_2 = t_3$ by the amount

$$\sqrt{v^2 + v^2 + v^2} = v\sqrt{3}$$

(see diagram).

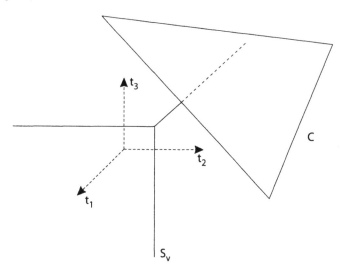

A convenient device for graphing both sets in two dimensions and for removing the effect of the translation at the same time is the use of *triangular coordinates*.

A visualization of the effect of these coordinates is obtained by imagining oneself as an observer in a position far out on the line $t_1 = t_2 = t_3$ and looking toward the origin. Then one would see the positive axes and some representative points as shown. The negative "octant" has been drawn in heavy lines.

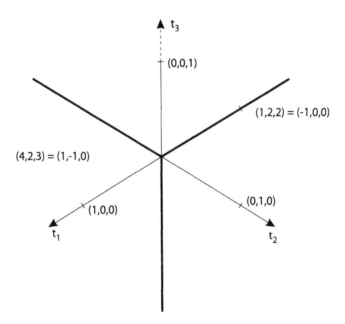

Remark that the points (t_1, t_2, t_3) and $(t_1 + a, t_2 + a, t_3 + a)$ cannot be distinguished in this projection. Thus the point (t_1, t_2, t_3) can be plotted by plotting $(t_1 - t_3, t_2 - t_3)$ in the (oblique) Cartesian system of coordinates in the plane, given by the projections of the t_1 and t_2 axes. The use of this system of coordinates is best illustrated by the example of the solution of a game.

EXAMPLE. Solve the matrix game

$$\begin{pmatrix} 3 & -2 & 4 \\ -1 & 4 & 2 \\ 2 & 2 & 6 \end{pmatrix}.$$

SOLUTION. We graph the octant S_v (indeed *all* octants) in heavy lines and the convex set C, which in this case is the triangle spanned by the three points $(3, -1, 2)$, $(-2, 4, 2)$ and $(4, 2, 6)$.

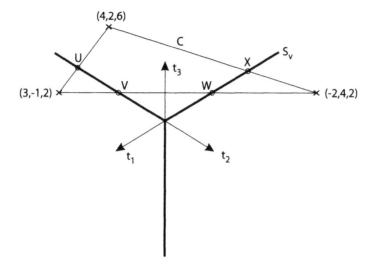

It is visually obvious in the figure that the contact between C and S_v occurs on the polygon (which may degenerate to a line segment or a point) and that the only possibilities for the vertices of this polygon are the three vertices of C and the circled points U, V, W, and X. Just as in the proof of the fundamental theorem, we determine the octant S_e that is in contact with a point $T = (t_1, t_2, t_3)$ by setting $e = E(T) = \max\{t_1, t_2, t_3\}$. Since player P_2 is trying to minimize his expected loss, the points under consideration that will appear in the contact with S_v are those which determine octants S_e with *minimal $e = v$*.

To carry out this program, we first must solve for the coordinates of the points U, V, W and X. The point $U = (u_1, u_2, u_3)$ is determined by the conditions

$$u_1 = u_3$$

$$(u_1, u_2, u_3) = a(3, -1, 2) + b(4, 2, 6)$$

for $a \geq 0$, $b \geq 0$, $a + b = 1$. These imply $3a + 4b = 2a + 6b$, that is, $a = 2b$, or finally $a = \frac{2}{3}$, $b = \frac{1}{3}$ and $U = (10/3, 0, 10/3)$. In a similar manner we obtain

$$U = \left(\frac{10}{3}, 0, \frac{10}{3}\right) \quad E(U) = \frac{10}{3}$$

$$V = (2, 0, 2) \quad\quad E(V) = 2$$

$$W = (0, 2, 2) \quad\quad E(W) = 2$$

$$X = \left(0, \frac{10}{3}, \frac{10}{3}\right) \quad E(X) = \frac{10}{3}$$

and for the vertices of C we have

$$\begin{aligned} E((3, -1, 2)) &= 3 \\ E((-2, 4, 2)) &= 4 \\ E((4, 2, 6)) &= 6. \end{aligned}$$

Hence the value of the game is 2, the minimum value of $E(T)$ among the eligible T. The contact between S_2, and C occurs along the segment joining V to W. Since

$$V = \frac{4}{5}(3, -1, 2) \; + \; \frac{1}{5}(-2, 4, 2)$$

$$W = \frac{2}{5}(3, -1, 2) \; + \; \frac{3}{5}(-2, 4, 2)$$

player P_2 has two "extreme" optimal strategies

$$\bar{Y} = \begin{cases} \left(\dfrac{4}{5}, \dfrac{1}{5}, 0\right) \\[3mm] \left(\dfrac{2}{5}, \dfrac{3}{5}, 0\right) \end{cases}$$

and any optimal strategy for P_2 is a convex combination of these two.

Now that the nature of the contact is established, it is clear that the only separating hyperplane is the plane passing through $(2, 2, 2)$ and parallel to the $t_1 t_2$ plane, i.e.,

$$t_3 = 2.$$

Hence the unique optimal strategy for P_1 is the normal to this plane,

$$\bar{X} = (0, 0, 1).$$

EXERCISE 6. Solve the matrix game

$$\begin{pmatrix} 1 & -1 & 1 \\ 2 & 3 & 1 \end{pmatrix}.$$

EXERCISE 7. Solve the matrix game

$$\begin{pmatrix} 0 & 1 & 2 \\ 2 & 0 & 1 \\ 1 & 2 & 0 \end{pmatrix}.$$

2.7 An Algorithm for Solving All Matrix Games

Our aim in this section will be to convert the knowledge gained from the graphical method outlined in the previous section to construct an algorithm which will yield a solution to any matrix game in a finite (though possibly very large) number of steps [8].

A mixed strategy $X = (x_1, \ldots, x_m)$ for P_1 is an m-tuple of real numbers and hence lies quite naturally in R_m. The conditions $x_1 \geq 0$ for $i = 1, \ldots, m$ and $x_1 + \cdots + x_m = 1$ determine a set which appears frequently in mathematics and is called an $(m - 1)$-*dimensional simplex*. Thus a 1-dimensional simplex is a line segment, a 2-dimensional simplex is a triangle, while a 3-dimensional simplex is a tetrahedron.

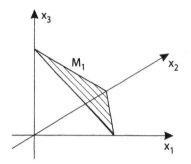
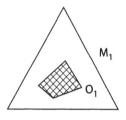

We will denote the set of mixed strategies for $P_1(P_2)$ by $M_1(M_2)$. The optimal strategies will form a certain subset $O_1(O_2)$ of $M_1(M_2)$. I contend that these subsets are (1) closed and (2) convex. These statements clearly need only be proved for one of the sets O_1, O_2.

(1) O_1 *is closed.* Suppose we have a mixed strategy X which is the limit of a sequence of optimal mixed strategies $X_1, X_2, \ldots$. Then $E(X_k, j) \geq v$ for $j = 1, \ldots, n$ and all k and the fact that E is a continuous function entail $E(X, j) \geq v$ for $j = 1, \ldots, n$ and thus X is an optimal strategy.

(2) O_1 *is convex.* Let $X = aX_1 + bX_2$ where X_1 and X_2 are optimal and $a \geq 0$, $b \geq 0$, $a + b = 1$. Then $E(X, j) = a\,E(X_1, j) + b\,E(X_2, j) \geq (a + b)v = v$ for all j and hence X is optimal.

In the 3×3 graphical example of Section 6 the solution set O_2 for P_2 looked like the diagram,

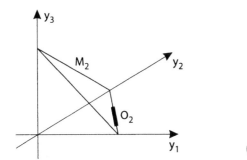

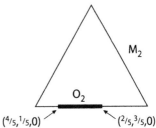

and it was possible to describe the entire set O_2 by giving its end points. Every optimal strategy for P_2 was a convex combination of these two points. It will be shown that this is true in general, that every optimal strategy is a convex combination of a finite number of extreme strategies.

DEFINITION 10. Let S be a convex set. A point U in S is called *extreme* if $U = (1/2)V + (1/2)W$ for no two distinct points V, W in S.

Examples

Example 1.
$S = \{T \mid t_1^2 + t_2^2 < 1\}$.
No extreme points.

(a)

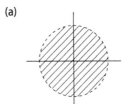

Example 2.
$S = \{T \mid t_1^2 + t_2^2 \leq 1\}$.
Extreme points are those T for which $t_1^2 + t_2^2 = 1$.

(b)

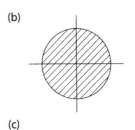

(c)

Example 3.
Extreme points are
U, V, W.

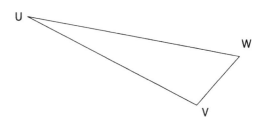

THEOREM 7. If C is a non-empty, closed, bounded, convex set in R_n then C is the convex hull of its extreme points.

PROOF. To prove this theorem it is necessary to prove an auxiliary lemma which is quite important in its own right.

DEFINITION 11. A point U is called a *boundary point* of a set S in R_n if $U \in S$ and U is not interior to S.

LEMMA. If U is a boundary point of a convex set C in R_n then there is a supporting hyperplane for C that passes through U.

PROOF. We may assume $U = 0$. Form $D = \{T \mid T = aV, \ V \in C, \ a \geqq 0\}$. Clearly D is convex. If we can show $D \neq R_n$ we will have the necessary hyperplane.

Assume $D = R_n$. Then we can choose n linearly independent vectors $V_1, \ldots, V_n$ in C. Moreover, since there are vectors in D in all directions from 0, for some a such that $1 > a > 0$, $-a(V_1 + \cdots + V_n) \in C$. Then the vectors $aV_1, \ldots, aV_n$ satisfy the hypothesis of Exercise 5 which implies that 0 is an interior point of C. But 0 is assumed to be a boundary point of C and this is a contradiction.

Therefore, by Theorem 4, there is a hyperplane $H(X, a)$ that supports D and hence C. Moreover $H(X, 0)$ is also such a support by the argument used at the end of the proof of Theorem 5 and passes through 0.

PROOF OF THEOREM 7. We prove the theorem by induction on n. It is clearly true for $n = 1$ (or even more clearly so for $n = 0$!). Assume that it holds in R_{n-1} and consider any boundary point U of C. By the lemma there is a supporting hyperplane H for C that passes through U. Then $C \cap H$ is a closed, bounded, convex set in H (which is an R_{n-1}) and hence the theorem holds for $C \cap H$. Therefore, boundary points are shown to be convex combinations of extreme points.

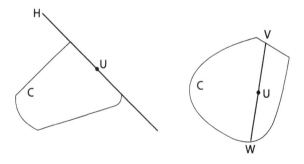

Consider any point U interior to C. Any line containing U must intersect C in a segment joining two boundary points V and W (the boundedness of C enters here) and $U = aV + bW$ where $a \geq 0$, $b \geq 0$ and $a + b = 1$. Hence, by taking the convex representations of V and W in terms of extreme points, we have the desired convex representation of U.

Remark that we can take one of V and W to be an extreme point and hence only one new point is introduced at each dimension. Therefore we can sharpen the statement to:

COROLLARY. Under the assumptions of Theorem 7, every point of C is a convex combination of at most $n + 1$ extreme points of C.

Thus we see that, in order to find all optimal strategies for a matrix game, it will suffice to find the extreme optimal strategies. Our plan for hunting these can be explained roughly in this manner. Equations being easier to handle than inequalities (the theory of linear equations being quite classical and well developed), we will try (successfully) to convert our search for extreme strategies into the problem of solving certain linear equations.

Suppose we know the value of the game to be v; then a point lies on the boundary of S_v only if some of its coordinates are equal to v and the rest are less than v. Hence, if $\bar{Y}$ is an optimal strategy,

$$a_{11}\bar{y}_1 + \cdots + a_{1n}\bar{y}_n = v$$

(1)
$$\cdots \quad \cdots \quad \cdots \quad \cdots$$

$$a_{r1}\bar{y}_1 + \cdots + a_{rn}\bar{y}_n = v$$

$$a_{r+1,1}\bar{y}_1 + \cdots + a_{r+1,n}\bar{y}_n < v$$

(2)
$$\cdots \quad \cdots \quad \cdots \quad \cdots$$

$$a_{m1}\bar{y}_1 + \cdots + a_{mn}\bar{y}_n < v$$

$$\bar{y}_j \geq 0 \quad \text{for } j = 1, \ldots, n$$

(3)
$$\bar{y}_1 + \cdots + \bar{y}_n = 1$$

where the rows of A may have to be renumbered to achieve this neat division of the inequalities into equations and strict inequalities. One very simple and obvious way to ensure that $\bar{Y}$ is an *extreme* optimal strategy is to assume that the equations:

$$a_{11}y_1 \quad + \cdots + \quad a_{1n}y_n \quad = \quad v$$

$$\cdots \qquad \cdots \qquad \cdots \qquad \cdots$$

(4)

$$a_{r1}y_1 \quad + \cdots + \quad a_{rn}y_n \quad = \quad v$$

$$y_1 \quad + \cdots + \quad y_n \quad = \quad 1$$

have the unique solution $y_1 = \bar{y}_1, \ldots, y_n = \bar{y}_n$. To prove this, assume $\bar{Y}$ is not extreme, i.e., $\bar{Y} = (1/2)Y' + (1/2)Y''$ with Y' and Y'' distinct optimal strategies. Then $E(i, Y') \leq v$ and $E(i, Y'') \leq v$ for all i but $(1/2)E(i, Y') + (1/2)E(i, Y'') = E(i, \bar{Y}) = v$ for $i = 1, \ldots, r$. Therefore, $E(i, Y') = E(i, Y'') = v$ for $i = 1, \ldots, r$ which would yield distinct solutions to (4). If we notice first that nothing in the proof demands that we include all of the rows i for which $E(i, \bar{Y}) = v$ in (4), and second that the proof remains valid as long as we include all the columns j for which $\bar{y}_j > 0$, then we have proved:

LEMMA 1. Let $\bar{Y} = (\bar{y}_1, \ldots, \bar{y}_n)$ be an optimal strategy for a matrix game with value v and $E(i, \bar{Y}) = v$ for $i = 1, \ldots, r$. If the system of equations

$$a_{11}y_1 \quad + \cdots + \quad a_{1s}y_s \quad = \quad v$$

$$\cdots \qquad \cdots \qquad \cdots \qquad \cdots$$

$$a_{r1}y_1 \quad + \cdots + \quad a_{rs}y_s \quad = \quad v$$

$$y_1 \quad + \cdots + \quad y_s \quad = \quad 1$$

has the unique solution $y_1 = \bar{y}_1, \ldots, y_s = \bar{y}_s$, then $\bar{Y}$ is an *extreme* optimal strategy.

Our object now will be to prove the converse of this lemma, that is, to construct such a system of equations for every extreme $\bar{Y}$. The reader should remark the close analogy presented by the equations which we construct and the equations solved in the graphical method of Section 6.

LEMMA 2. Let $\bar{Y} = (\bar{y}_1, \ldots, \bar{y}_n)$ be an optimal strategy for a matrix game A with value v. If $\bar{Y}$ is extreme then, by possibly renumbering the rows and columns of A, there exists an r and s such that the system of equations

$$a_{11}y_1 \quad + \cdots + \quad a_{1s}y_s - t = 0$$

$$\cdots \qquad \cdots \qquad \cdots \qquad \cdots$$

(5)

$$a_{r1}y_1 \quad + \cdots + \quad a_{rs}y_s - t = 0$$

$$y_1 \quad + \cdots + \quad y_s = 1$$

has the unique solution $y_1 = \bar{y}_1, \ldots, y_s = \bar{y}_s, t = v$.

PROOF. Form a system (5) by renumbering rows and columns, if and as necessary, so that $E(i, \bar{Y}) = v$ for $i = 1, \ldots, r$ and $E(i, \bar{Y}) < v$ for $i = r + 1, \ldots, m$, and $\bar{y}_j > 0$ for $j = 1, \ldots, s$ and $\bar{y}_j = 0$ for $j = s + 1, \ldots, n$. Clearly, $r \geqq 1$, otherwise v would not be the value of the game; clearly $s \geq 1$ since $\bar{y}_1 + \cdots + \bar{y}_n = 1$. I contend that the system (5) so obtained has the desired properties. For, if (5) has two distinct solutions $y'_1, \ldots, y'_s, t'$ and $y''_1, \ldots, y''_s, t''$, consider the n-dimensional vectors

$$\bar{Y}' = (\bar{y}_1 + \epsilon(y'_1 - y''_1), \ldots, \bar{y}_s + \epsilon(y'_s - y''_s), 0, \ldots, 0)$$

$$\bar{Y}'' = (\bar{y}_1 - \epsilon(y'_1 - y''_1), \ldots, \bar{y}_s - \epsilon(y'_s - y''_s), 0, \ldots, 0).$$

Since $\sum_j (y'_j - y''_j) = 1 - 1 = 0$, Y' and Y'' have component sums equal to one; and, if we choose $\epsilon > 0$ small enough, these components are non-negative since $\bar{y}_j > 0$ for $j = 1, \ldots, s$. Now assume $t' \leqq t''$ and we will verify that Y' is an optimal strategy. There are two cases:

CASE 1. Rows $i = 1, \ldots, r$. Here $E(i, Y') = v + \epsilon(t' - t'') \leqq v$.

CASE 2. Rows $i = r + 1, \ldots, m$. Here

$$E(i, Y') = E(i, \bar{Y}) + \epsilon \sum_j a_{ij}(y'_j - y''_j) < v$$

for $\epsilon > 0$ chosen small enough.

Hence, Y' is a solution, and moreover $t' = t''$ since otherwise $E(i, Y')$ would be less than v for all i and hence the value of the game would be less than v. Therefore, the same two cases show that Y'' is an optimal strategy, and hence we have constructed two distinct optimal strategies. However, direct computation shows that $\bar{Y} = (1/2)Y' + (1/2)Y''$. This contradicts the assumption that $\bar{Y}$ is extreme and completes the proof of the lemma.

It is well known [9] that the system of equation (5) has a unique solution if and only if the corresponding homogeneous system has only the trivial solution $y_1 = \ldots = y_s = t = 0$. This is equivalent to the statement that the *columns* of the matrix of coefficients:

$$\begin{pmatrix} a_{11} & \cdots & a_{1s} & -1 \\ \vdots & & \vdots & \vdots \\ a_{r1} & \cdots & a_{rs} & -1 \\ 1 & \cdots & 1 & 0 \end{pmatrix}$$

are linearly independent. But a matrix has just as many linearly independent rows as columns [9] and so we can assert that we can choose the last row and s rows from among the first r rows of coefficients such that:

$$(6) \qquad K = \begin{pmatrix} a_{11} & \cdots & a_{1s} & -1 \\ \vdots & & \vdots & \vdots \\ a_{s1} & \cdots & a_{ss} & -1 \\ 1 & \cdots & 1 & 0 \end{pmatrix}$$

has linearly independent rows and columns (the rows may have to be renumbered again). We shall call the matrix

(7)
$$K_0 = \begin{pmatrix} a_{11} & \cdots & a_{1s} \\ \vdots & & \vdots \\ a_{s1} & \cdots & a_{ss}, \end{pmatrix}$$

which is a square submatrix of the game matrix, a *kernel* for the extreme optimal strategy (not "the" kernel since the steps in its construction were not unique). Now that we have the non-singular matrix K, we can compute the solution of (5) by the usual methods of determinants [9]. Denote the determinants of K and K_0 by $|K|$ and $|K_0|$ respectively and the co-factor of a_{ij} in K_0 by K_{ij}. Then, expanding K by its last row and then by its last column, we have

$$|K| = \sum_{i,j} K_{ij} \neq 0 \quad \text{since } K \text{ is non-singular.}$$

The solution of (5) is then

$$y_j = \sum_i K_{ij} \Big/ |K| = \sum_i K_{ij} \Big/ \sum_{i,j} K_{ij}$$

$$t = |K_0| \Big/ |K| = |K_0| \Big/ \sum_{i,j} K_{ij}.$$

We can rephrase these calculations, combined with Lemma 2 in the following manner:

THEOREM 8. Let $\bar{Y} = (\bar{y}_1, \ldots, \bar{y}_n)$ be an optimal strategy for a matrix game A with value v. If $\bar{Y}$ is extreme then there is a square submatrix K_0 of A such that

(8)
$$\bar{y}_j = \sum_i K_{ij} \Big/ \sum_{i,j} K_{ij}$$

$$v = |K_0| \Big/ \sum_{i,j} K_{ij}$$

where K_{ij} is the co-factor of a_{ij} in K_0, and i and j range over the rows and columns of K_0.

Clearly, an exactly analogous theorem holds for optimal strategies $\bar{X}$ for the first player. This theorem provides the basis for the following algorithm.

ALGORITHM. Given a matrix game A, examine all square submatrices K_0 by computing *possible* extreme optimal strategies $\bar{X}$ and $\bar{Y}$ and *possible* values v by (7). Those $\bar{X}$ and $\bar{Y}$ which are optimal strategies with correct v calculated by (7) constitute all extreme optimal strategies. All optimal strategies are convex combinations of the finite number of extreme optimal strategies obtained in this manner.

PROOF. Theorem 8 assures us that all extreme optimal strategies will be found in this manner. There are three reasons why a proposed K_0 may not provide an extreme optimal strategy.

(1) The $\bar{X}$ or $\bar{Y}$ as calculated by (7) may not be mixed strategies.
(2) The $\bar{X}$ or $\bar{Y}$ as calculated by (7) may not be optimal strategies.
(3) The v as calculated by (7) might not be the true value of the game.
   The following example illustrates all three types of failure.

EXAMPLE. Solve the matrix game

$$\begin{pmatrix} 3 & -2 & 4 \\ -1 & 4 & 2 \\ 2 & 2 & 6 \end{pmatrix}$$

SOLUTION. The only 3×3 kernel consists of the game itself, i.e.,

$$K_0 = \begin{pmatrix} 3 & -2 & 4 \\ -1 & 4 & 2 \\ 2 & 2 & 6 \end{pmatrix}$$

The matrix of cofactors (K_{ij}) is easily calculated to be

$$\begin{pmatrix} 20 & 10 & -10 \\ 20 & 10 & -10 \\ -20 & -10 & 10 \end{pmatrix}$$

and hence, using (7), $\bar{X} = (1, 1, -1)$, $\bar{Y} = (1, 1/2, -1/2)$ and K_0 can be discarded because these are not even mixed strategies. The 2×2 kernels

$$\begin{pmatrix} 3 & 4 \\ -1 & 2 \end{pmatrix}, \begin{pmatrix} -2 & 2 \\ 2 & 6 \end{pmatrix}, \begin{pmatrix} -2 & 4 \\ 2 & 6 \end{pmatrix}$$

can be discarded for the same reason. All of the other 2×2 kernels yield a mixed strategy for one or both players. When we examine the kernel

$$\begin{pmatrix} 3 & -2 \\ -1 & 4 \end{pmatrix}$$

we find $\bar{X} = (1/2, 1/2, 0)$, $\bar{Y} = (3/5, 2/5, 0)$ and $v = 1$. Here $\min_j E(\bar{X}, j) = 1$ and $\max_i E(\bar{Y}, i) = 2$. Hence $\bar{X}$ is an extreme optimal strategy if it is an optimal strategy but $\bar{Y}$ is surely not extreme since its value does not agree with v. However, when we consider the 2×2 kernels

$$\begin{pmatrix} 3 & -1 \\ 2 & 2 \end{pmatrix} \quad \text{and} \quad \begin{pmatrix} -1 & 4 \\ 2 & 2 \end{pmatrix}$$

we find $\bar{X} = (0, 0, 1)$, $\bar{Y} = (4/5, 1/5, 0)$, $v = 2$ and $\bar{X} = (0, 0, 1)$, $\bar{Y} = (2/5, 3/5, 0)$, $v = 2$ respectively. These are fully satisfactory since $\max_j E(\bar{X}, j) = \min_i E(i, \bar{Y}) = 2$. and hence the $\bar{X}$ previously considered was not optimal. The remaining 2×2 kernels

$$\begin{pmatrix} 3 & 4 \\ 2 & 6 \end{pmatrix}, \begin{pmatrix} -2 & 4 \\ 4 & 2 \end{pmatrix}, \begin{pmatrix} 4 & 2 \\ 2 & 6 \end{pmatrix}$$

can be excluded since they do not yield the proper value for the game. As for 1×1 kernels, now that we know the value of the game is 2, we need only hunt for the entry 2 in the matrix and check whether it is the maximum in its column or the minimum in its row (that is, whether the corresponding $\bar{X}$ and $\bar{Y}$ are optimal). There are two such entries in the third row and they provide $\bar{X} = (0, 0, 1)$ again.

Summarizing, the extreme optimal strategies are

$$\bar{X} = (0, 0, 1)$$

$$\bar{Y} = \begin{cases} \left(\dfrac{4}{5}, \dfrac{1}{5}, 0\right) \\ \\ \left(\dfrac{2}{5}, \dfrac{3}{5}, 0\right) \end{cases}$$

and the value of the game is 2. The reader should compare this solution with the graphical solution of Section 6 and connect our algebraic reasons for discarding kernels with the figure given there.

EXERCISE 8. Show that the matrix game

$$\begin{pmatrix} a & 0 & 0 \\ 0 & b & 0 \\ 0 & 0 & c \end{pmatrix}$$

where $a > 0$, $b > 0$, and $c > 0$, has a unique solution and find it.

2.8 Simplified Poker

In this section we shall apply the techniques developed thus far to the solution of a rather large matrix game patterned on the familiar game of Poker [10]. As actually played, Poker is far too complex a game to permit a complete analysis at present; however, this complexity is computational and the restrictions that we shall impose serve only to bring the numbers involved within a reasonable range. The only restriction that is not of this nature consists in setting the number of players at two. The simplifications, though radical, enable us to compute *all* optimal strategies for both players. In spite of these modifications, it seems that Simplified Poker retains many of the essential characteristics of the usual game. This will be our first encounter with a game given not in matrix form but by its rules, and our immediate problem will be to reduce the verbal description to a payoff matrix.

SIMPLIFIED POKER. An *ante* of one unit is required of each of the two players. They obtain a fixed *hand* at the beginning of a play by drawing one card apiece from a pack of three cards (rather than the $\binom{52}{5} = 2{,}598{,}960$ hands possible in five card Stud Poker) numbered 1,2,3. Then the players choose alternatively to *bet* one unit or to *pass* without betting. Two successive bets or passes terminate a play, at which time the player holding the higher card wins the amount wagered previously by the other player. A player passing after a bet also ends a play and loses his ante.

Thus thirty possible plays are permitted by the rules. First of all, there are six possible deals; for each deal the action of the players may follow one of five courses which are described in the following diagram:

	First Round Player P_1	Second Round Player P_2	Player P_1	Payoff
Deal	Pass	Pass		1 to holder of higher card
		Bet	Pass	1 to P_2
			Bet	2 to holder of higher card
	Bet	Pass		1 to P_1
		Bet		2 to holder of higher card

What are the pure strategies available to the two players in this game? By our interpretation of matrix games, a pure strategy is a mode of action decided upon before a play of the game such that the choice of a pure strategy for each of the players determines the play uniquely. Now this certainly entails making a choice for each of the possible situations that might confront a player. For P_1 this means that he must choose whether to pass or bet for each of the three deals possible to him. Moreover, if he has decided to pass in the first round and wants to prepare for all possible contingencies, he must decide whether he will pass or bet in the second round. An example of such a plan is: Bet in the first round when dealt a 1, always pass with a 2, and wait until the second round to bet on a 3. A convenient way to code these plans is by ordered triples (a_1, a_2, a_3) where a_k gives the instructions for card $k = 1, 2, 3$. Then a_k is deciphered by expanding it in the binary system, the first figure giving instructions for the first round of betting, the second giving instruction for the second round, with 0 meaning pass and 1 meaning bet. Thus

$$a_k = \begin{cases} 0 &= 00 \\ 1 &= 01 \\ 2 &= 10 \end{cases} \text{ means } \begin{cases} \text{pass in round 1; pass in round 2} \\ \text{pass in round 1; bet in round 2} \\ \text{bet in round 1 .} \end{cases}$$

In this code the plan described above becomes $(a_1, a_2, a_3) = (2, 0, 1)$.

A similar system of coding can be used for the pure strategies available to P_2. Here we use triples (b_1, b_2, b_3) where $b_\ell = 0, 1, 2, 3$ gives instructions for card ℓ when expanded in the binary system. The first figure gives directions when confronted by a pass, the second when confronted by a bet, again with 0 meaning pass and 1 meaning bet. Thus $(b_1, b_2, b_2) = (2, 0, 1) = (10, 00, 01)$ means that P_2 should pass except when holding a 1 and confronted by a pass or when holding a 3 and confronted by a bet.

In terms of this description of pure strategies, the payoff to P_1 is given by the following scheme (*this is not the game matrix yet!*):

a_k $\quad$ b_l	$0 = 00$	$1 = 01$	$2 = 10$	$3 = 11$
$0 = 00$	± 1	± 1	-1	-1
$1 = 01$	± 1	± 1	± 2	± 2
$2 = 10$	1	± 2	1	± 2

where the ambiguous sign is $+$ if $k > \ell$ and is $-$ if $k < \ell$. From the coding of the pure strategies it is clear that P_1 has $3 \times 3 \times 3 = 27$ pure strategies while P_2 has $4 \times 4 \times 4 = 64$ pure strategies. Before blindly proceeding to calculate the 27×64 payoff matrix it is prudent to eliminate a number of these strategies by the application of some Poker sense. For instance, P_1 would certainly be unwise to bet with a 1 in the second round since, upon comparing cards, he would certainly lose 2 instead of the 1 he would lose if he passed. This verbal argument corresponds to the fact that all the matrix entries in the row $(1, a_2, a_3)$ are less than or equal to those in the row $(0, a_2, a_3)$. The method that we are using implicitly here is of such general applicability that it justifies a digression to put it in its proper context.

DEFINITION 12: Let i_1 and i_2 be distinct pure strategies for P_1 in an $m \times n$ game matrix $A = (a_{ij})$. Then i_1 is said to *dominate* i_2 if

$$a_{i_1 j} \geqq a_{i_2 j} \quad \text{for } j = 1, \ldots, n.$$

Similarly, if j_1 and j_2 are distinct pure strategies for P_2, j_1 *dominates* j_2 if

$$a_{ij_j} \leqq a_{a_{ij_2}} \quad \text{for } i = 1, \ldots, m.$$

PROPOSITION 3. Let $A = (a_{ij})$ be an $m \times n$ matrix game in which row 2 dominates row 1. Let B be the matrix game

$$\begin{pmatrix} a_{21} & \cdots & a_{2n} \\ \vdots & & \vdots \\ a_{m1} & \cdots & a_{mn} \end{pmatrix}.$$

Then the value of B is equal to the value of A and if the mixed strategy $(\bar{x}_2, \ldots, \bar{x}_m)$ is optimal for B then $\bar{X} = (0, \bar{x}_2, \ldots, \bar{x}_m)$ is optimal for A.

PROOF. Let $\bar{X} = (\bar{x}_2, \ldots, \bar{x}_m)$, $\bar{Y} = (\bar{y}_1, \ldots, \bar{y}_n)$, and v be a solution for B, that is,

$$a_{2j}\bar{x}_2 + \cdots + a_{mj}\bar{x}_m \geq v \quad \text{for } j = 1, \ldots, n$$

$$a_{i1}\bar{y}_1 + \cdots + a_{in}\bar{y}_n \leq v \quad \text{for } i = 2, \ldots, m.$$

Then

$$E(\bar{X}, j) = a_{2j}\bar{x}_2 + \cdots + a_{mj}\bar{x}_m \geq v \quad \text{for } j = 1, \ldots, n$$

$$E(i, \bar{Y}) = a_{i1}\bar{y}_1 + \cdots + a_{in}\bar{y}_n \leq v \quad \text{for } i = 1, \ldots, m$$

and

$$E(1, \bar{Y}) = a_{11}\bar{y}_1 + \cdots + a_{1n}\bar{y}_n \leq a_{21}\bar{y}_1 + \cdots + a_{2n}\bar{y}_n \leq v$$

hence $\bar{X}$, $\bar{Y}$, and v are a solution for the game A.

Of course the analogous theorem holds for any pair of rows and columns. The content of these theorems is that it is possible to solve a game in which a row or column is dominated by deleting it, solving the new game, and extending the optimal strategies by zeros in the deleted rows and columns. It should be remarked, however, that we may lose solutions to the original game with this procedure. Consider as an example of this phenomenon the matrix game

$$\begin{pmatrix} 0 & 0 \\ -1 & 1 \end{pmatrix}$$

in which it is clear that column 1 dominates column 2 and hence we may solve

$$\begin{pmatrix} 0 \\ -1 \end{pmatrix}$$

and extend the unique solution obtained to the solution $\bar{X} = (1, 0)$, $\bar{Y}(1, 0)$, and $v = 0$ for the original matrix. However, the original game has another extreme optimal strategy for P_2, $\bar{Y} = (1/2, 1/2)$ which obviously cannot be obtained by this procedure. Of course, one might raise the objection: why should P_2 risk paying 1 by playing his second pure strategy when he risks nothing at all by playing his first pure strategy? The answer is that this is a consequence of our acceptance of the pessimistic expectation as a basis of the concept of solution, and without more subtle distinctions between optimal strategies [11] we must consider $\bar{Y} = (1, 0)$ and $\bar{Y} = (1/2, 1/2)$ as equally good since, for both, $\max_i E(i, \bar{Y}) = 0$.

Returning now to Simplified Poker, we remark that rows and columns that tell P_1 to bet on a 1 or P_2 to pass on a 3 when confronted by a bet are dominated and hence may be dropped from consideration. Now that these strategies have been eliminated new dominations appear. First, we notice that if P_1 holds a 2 he may as well pass in the first round, deciding to bet in the second if confronted by a bet, as bet originally. On either strategy he will lose the same amount if P_2 holds a 3; on the other hand, P_2 may bet on a 1 if confronted by a pass but certainly will not if confronted by a bet. Second, P_2 may as well pass as bet when holding a 2 and confronted by a pass, since P_1 will now answer a bet only when he holds a 3. (It may be remarked that the Skin Game of Section 1 provides a good example of the successive appearance of dominations. After we have eliminated row 1

$$\begin{pmatrix} \cancel{-1} & \cancel{-1} & \cancel{-2} \\ -1 & 1 & 1 \\ 2 & -1 & 0 \end{pmatrix}$$

it is seen that column 3 is now dominated by column 2 and we are left with a 2×2 matrix game to solve.

$$\begin{pmatrix} \cancel{-1} & \cancel{-1} & \cancel{-2} \\ -1 & 1 & \cancel{1} \\ 2 & -1 & \cancel{0} \end{pmatrix}$$

Furthermore, after we have solved this 2×2 game, an application of Proposition 2 assures us that the deleted strategies do not appear in any optimal strategy for the full game. (The careful solver of Simplified Poker who may wish to find all of the optimal strategies will have to make complementary arguments of the same nature to verify that none of the pure strategies that we have just dropped can appear in an optimal strategy with positive probability.)

We are now in a position to compute the game matrix composed of those strategies not eliminated by the previous domination arguments. To illustrate these computations suppose P_1 chooses his pure strategy $(0, 1, 2)$ and P_2 his pure strategy $(2, 1, 3)$. Then we must calculate the payoff for each possible deal (k, ℓ) of card k to P_1 and card ℓ to P_2.

Deal	Payoff	Deal	Payoff
(1,2)	-1	(2,3)	-2
(1,3)	-1	(3,1)	$+1$
(2,1)	$+2$	(3,2)	$+2$

Hence the expected payoff by P_2 to P_1 is

$$\frac{1}{6}(-1) + \frac{1}{6}(-1) + \frac{1}{6}(2) + \frac{1}{6}(-2) + \frac{1}{6}(1) + \frac{1}{6}(2) = \frac{1}{6}.$$

Thus, the choice of two pure strategies does not determine a play uniquely but determines a set of plays with fixed probabilities and the matrix entry that we compute is an expectation. The final result of this calculation for all undominated pure strategies is:

(a_1, a_2, a_3) \ (b_1, b_2, b_3)	$(0, 0, 3)$	$(0, 1, 3)$	$(2, 0, 3)$	$(2, 1, 3)$
$(0, 0, 1)$	0	0	$-\dfrac{1}{6}$	$-\dfrac{1}{6}$
$(0, 0, 2)$	0	$\dfrac{1}{6}$	$-\dfrac{1}{3}$	$-\dfrac{1}{6}$
$(0, 1, 1)$	$-\dfrac{1}{6}$	$-\dfrac{1}{6}$	$\dfrac{1}{6}$	$\dfrac{1}{6}$
$(0, 1, 2)$	$-\dfrac{1}{6}$	0	0	$\dfrac{1}{6}$
$(2, 0, 1)$	$\dfrac{1}{6}$	$-\dfrac{1}{3}$	0	$-\dfrac{1}{2}$
$(2, 0, 2)$	$\dfrac{1}{6}$	$-\dfrac{1}{6}$	$-\dfrac{1}{6}$	$-\dfrac{1}{2}$
$(2, 1, 1)$	0	$-\dfrac{1}{2}$	$\dfrac{1}{3}$	$-\dfrac{1}{6}$
$(2, 1, 2)$	0	$-\dfrac{1}{3}$	$\dfrac{1}{6}$	$-\dfrac{1}{6}$

When we realize that even this reduced payoff matrix has 168 2×2 submatrices, 224 3×3 submatrices, and 70 4×4 submatrices it becomes clear that our algorithm should not be applied without further thought. However, let us take two kernels at random, say

$$\begin{pmatrix} 0 & 0 & -\dfrac{1}{6} \\ 0 & \dfrac{1}{6} & -\dfrac{1}{3} \\ -\dfrac{1}{6} & -\dfrac{1}{6} & \dfrac{1}{6} \end{pmatrix} \quad \text{and} \quad \begin{pmatrix} 0 & -\dfrac{1}{6} \\ -\dfrac{1}{6} & \dfrac{1}{6} \end{pmatrix}$$

which are respectively the first three rows and columns and the first and third rows combined with the first and fourth columns. These yield

$$\bar{X} = \left(\frac{2}{3}, 0, \frac{1}{3}, 0, 0, 0, 0, 0\right)$$

$$\bar{Y} = \left(\frac{1}{3}, \frac{1}{3}, \frac{1}{3}, 0\right)$$

$$v = -\frac{1}{18}$$

and

$$\bar{X} = \left(\frac{2}{3}, 0, \frac{1}{3}, 0, 0, 0, 0, 0\right)$$

$$\bar{Y} = \left(\frac{2}{3}, 0, 0, \frac{1}{3}\right)$$

$$v = -\frac{1}{18} \quad .$$

It can be verified immediately that these are optimal strategies with the given value and hence are extreme optimal strategies. Now we can provide an example of the complementary arguments that assure us that no solutions are lost by discarding dominated rows and columns. Consider the pure strategies of the form $(1, a_2, a_3)$ which we have eliminated for P_1. Our verbal argument showed that he does at least as well using $(0, a_2, a_3)$ no matter what mixed strategy P_2 uses and irrespective of the deal. However, both of the optimal $\bar{Y}$ given above instruct P_2 always to bet on a 3, and hence for the deal $(1,3)$ the loss incurred by P_1 will be one unit for $(0, a_2, a_3)$ and two units for $(1, a_2, a_3)$. Since this deal occurs 1/6 of the time

$$E\left((1, a_2, a_3), \bar{Y}\right) = E\left((0, a_2, a_3), \bar{Y}\right) - \frac{1}{6} < -\frac{1}{18}$$

and Proposition 2 tells us that P_1 cannot play $(1, a_2, a_3)$ in any optimal strategy. The other dominated strategies are handled by similar arguments.

Further examination of the kernels yields the following list of optimal strategies:

$$\bar{X} = \left(\frac{2}{3}, 0, \frac{1}{3}, 0, 0, 0, 0, 0\right)$$

$$\left(\frac{1}{3}, 0, 0, \frac{1}{2}, 0, \frac{1}{6}, 0, 0\right)$$

$$\left(\frac{5}{9}, 0, 0, \frac{1}{3}, 0, 0, \frac{1}{9}, 0\right)$$

$$\left(\frac{1}{2}, 0, 0, \frac{1}{3}, 0, 0, 0, \frac{1}{6}\right)$$

$$\left(0, \frac{2}{5}, \frac{7}{15}, 0, \frac{2}{15}, 0, 0, 0\right)$$

$$\left(0, \frac{1}{3}, \frac{1}{2}, 0, 0, \frac{1}{6}, 0, 0\right)$$

$$\left(0, \frac{1}{2}, \frac{1}{3}, 0, 0, 0, \frac{1}{6}, 0\right)$$

$$\left(0, \frac{4}{9}, \frac{1}{3}, 0, 0, 0, \frac{2}{9}\right)$$

$$\left(0, \frac{1}{6}, 0, \frac{7}{12}, \frac{1}{4}, 0, 0, 0\right)$$

$$\left(0, \frac{5}{12}, 0, \frac{1}{3}, 0, 0, \frac{1}{4}, 0\right)$$

$$\left(0, \frac{1}{3}, 0, \frac{1}{3}, 0, 0, 0, \frac{1}{3}\right)$$

$$\left(0, 0, 0, \frac{2}{3}, 0, \frac{1}{3}, 0, 0\right)$$

$$\bar{Y} = \left(\frac{1}{3}, \frac{1}{3}, \frac{1}{3}, 0\right)$$

$$\left(\frac{2}{3}, 0, 0, \frac{1}{3}\right) .$$

The problem of showing that these constitute all extreme optimal strategies is a difficult one (other than by the direct examination of the 486 kernels): however, we can now show that the set is complete for P_2. To do this, let $A = (a_{ij})$ be the 4×8 payoff matrix given above and let $B = (b_{ij})$ where $b_{ij} = a_{ij} + 1/18$ for all i and j. Then, by Proposition 1, the game defined by B has value zero and the same optimal strategies as A. Moreover, since all the pure strategies for P_1 appear with positive probability in some optimal $\bar{X}$ above, Proposition 2 implies that a mixed strategy $Y = (y_1, y_2, y_3, y_4)$ is optimal if and only if

$$b_{i1}y_1 + b_{12}y_2 + b_{i3}y_3 + b_{i4}y_4 = 0 \text{ for } i = i, \ldots, 8$$

that is, Y lies in the null-space of B [9]. We have given two distinct such Y above and hence the column nullity of B is at least two. On the other hand, every pair of columns of B is linearly independent and hence the column rank is at least two. But

Column rank + Column nullity = Number of columns

and hence for B,

Column rank = Column nullity = 2.

Therefore the set of optimal strategies for P_2 is contained in the intersection of the 2-dimensional space spanned by the $\bar{Y}$ given above with the hyperplane $y_1 + \cdots + y_4 = 1$ and is therefore 1-dimensional. Hence the two given extreme strategies are the only extreme strategies for P_2. (This line of reasoning has been applied to arbitrary games to give quite elegant characterizations of the possible solution sets for an $m \times n$ matrix game [12].)

A striking simplification of the solution is achieved if we return to the verbal description of Simplified Poker. We do this by introducing *behavior strategies* to describe the choices remaining available to the players after we have eliminated the dominated strategies. We define for

P_1: $\alpha = $ probability of bet with 1 in first round
 $\beta = $ probability of bet with 2 in second round
 $\gamma = $ probability of pass with 3 in first round

P_2: $\xi = $ probability of bet with 1 against a pass
 $\eta = $ probability of bet with 2 against a bet.

In terms of behavior strategies, P_1's optimal strategies are found to lie on the segment

$$\alpha = \frac{\gamma}{3}$$

$$\beta = \frac{\gamma}{3} + \frac{1}{3}$$

$$0 \leqq \gamma \leqq 1.$$

These may be described verbally by saying that P_1 may pass on a 3 in the first round with arbitrary probability, but then he must bet on a 1 in the first round one-third as often, while the probability that he bets on a 2 in the second round is one-third more than the probability with which he bets on a 1 in the first round.

On the other hand, we find that P_2 has the unique optimal behavior strategy

$$(\xi, \eta) = \left(\frac{1}{3}, \frac{1}{3}\right),$$

which instructs him to bet one-third of the time when holding a 1 and confronted by a pass and to bet one-third of the time when holding a 2 and confronted by a bet.

The extreme simplicity of this description is no accident; in the next chapter we will investigate the exact conditions under which it is possible and give an alternative solution of this game.

The presence of *bluffing* and *underbidding* in these solutions is noteworthy (bluffing means betting with a 1; underbidding means passing with a 3). All but the extreme behavior strategies for P_1 include both buffing and underbidding while the single behavior strategy for P_2 instructs him to bluff one-third of the time (underbidding is not available to him).

Notes

1. This is the first place in the theory that mathematical expectation enters. For the mathematical theory, it is a definition of what we are going to maximize. For the interpreted theory, some justification must be make for its use. It is calculated in the following manner: If the mutually exclusive events with numerical values $a_1, \ldots, a_m$ occur with probabilities $p_1, \ldots, p_m$ then the expected value is $p_1 a_1 + \cdots + p_m a_m$. It is possible to make an argument for other criteria in games such as is described by the following matrix:

$$\begin{pmatrix} 0 & 0 \\ -1 & 1,000,000 \end{pmatrix}.$$

Evidently, P_2 should never play his second column, since to do so is to expose himself to a tremendous loss. However, his expected loss is clearly zero because P_1 has a strategy which gives him exactly this no matter what he does while P_2 can protect himself against losing more by playing his first column all the time. But the strange thing is he can also achieve this expectation by playing his first column with probability 1,000,000/1,000,001 and his second column with probability 1/1,000,001. One can imagine interpretations in which this is risky!

2. An amusing story was built around this example. See E. S. Locke, "The finan-seer," *Astounding Science Fiction*, **135–136** (Oct. 1949).

3. See J. von Neumann, "Zur Theorie der Gesellshaftsspiele," *Math. Annalen*, **100** (1928), 295-300. A discussion of the history of this theorem in the literature is given in the Appendix of this chapter.

4. The most complete survey of the subject of convex sets was made by T. Bonnesen and W. Fenchel, *Theorie der Konvexen Körper*, Ergebnisse der Mathematik und ihrer Grenzgbiete, vol. 3, no. 1 (Julius Springer, Berlin, 1934; reprint Chelsea, New York, 1948). However, for the selection of topics of interest in game theory the reader may find the following references to be of use:

Caratheodory, C., "Über den Variabilitätsbereich der Fourierscher: Konstanten ...," *Rend. Circ. mat. Palermo*, **32** (1911), 193–217.

Fenchel, W. "Uber Krümmung und Windung geschlossener Raumkurven," *Math. Annalen*, **101** (1929), 238–252.

Weyl, H., "Elementare Theorie der konvexen Polyder," *Comm. Math. Helv.*, **7** (1934–35), 290–306 (English translation: "The Elementary Theory of Convex Polyhedra," in *Annals of Math. Study* No. 24, Princeton, 1950).

Gale, D., "Convex polyhedral cones and linear inequalities" in *Activity Analysis of Production and Allocation*, edited by T. C. Koopmanns, John Wiley & Sons, New York, 1951.

5. This theorem was certainly known to Minkowski as early as 1896. See, for example § 16 in H. Minkowski, *Geometrie der Zahlen*, B. G. Teubner, Leipzig and Berlin, 1910.

A neglected algebraic proof that can scarcely be improved upon now was given by Farkas in 1902 for the case of polyhedral convex sets R_n (i.e., sets which are the intersection of a finite number of half-spaces): J. Farkas, "Theorie der einfachen Ungleichungen," *J. Reine Angew. Math.*, **124** (1902), 1–27.

6. By a *closed* set $S \subset R_n$, we mean a set which contains all points X such that, given $\epsilon > 0$, there exists $X_\epsilon \in S$ and $\| X - X_\epsilon \| < \epsilon$. Such a point X is called a *limit point* of S. By choosing a sequence $\epsilon_1, \epsilon_2, \ldots$ of positive ϵ_k which tend to zero and denoting the corresponding points of S by X_k, we see that X is a *limit point* of S if and only if there is a sequence $X_1, X_2, \ldots$ of points $X_k \in S$ such that $\| X - X_k \|$ tends to zero as k tends to ∞. We will sometimes write this as $X_k \longrightarrow X$. Thus, 0 is a limit point of the positive rational numbers but is not positive and hence the set of positive rational numbers is not closed.

7. For an introduction to the pathology of convex sets in infinite dimensional vector spaces, see J. Tukey, "Some notes on the separation of convex sets," *Portugaliae Math.* **3** (1942), 95–102.

The problem posed there has since been settled by J. Dieudonné, "Sur la separation des ensembles convèxes dans un espace de Banach," *Revue Sci. (Rev. Rose Illus.)*, **81** (1943), 277–278.

8. The first step in the solution of this problem was made by I. Kaplansky, "A contribution to von Neumann's theory of games," *Ann. of Math.* **2/46** (1945),

474–479, who gave an inductive procedure for determining the value of an arbitrary matrix game in a finite number of steps. The problem was completed by L. S. Shapley and R. N. Snow, "Basic solutions of discrete games" in *Annals of Math. Study* No. 24, Princeton, 1950.

Also related to this problem are the results of T. S. Motzkin, H. Raiffa, G. L. Thompson, and R. M. Thrall, "The double description method" in Annals of Math. Study No. 28, Princeton, 1952.

9. Concise and modern treatments of the important facts about matrices and vectors can be found in:

Artin, E., *Galois Theory*, 2nd ed., Notre Dame Math. Lectures No. 2, 1944.

Birkhoff, G., and Maclane, S., *A Survey of Modern Algebra*, Macmillan, New York, 1946.

Halmos, P. R., "Finite dimensional vector spaces," Annals of Math. Study No. 7, Princeton, 1942.

10. A fuller treatment of this game can be found in H. W. Kuhn, "Simplified Poker" in Annals of Math. Study No. 24. This paper also contains references to previous models of Poker.

11. One such notion is that of "permanent optimality," introduced by von Neumann and Morgenstern.

12. See the following references:

Gale, D. and Sherman, S., "Solutions of finite two-person games" in Annals of Math. Study No. 24.

Bohnenblust, H. F., Karlin, S., and Shapley, L. S., "Solutions of discrete two-person games" in Annals of Math. Study No. 24.

APPENDIX

THE FUNDAMENTAL THEOREM

In Chapter 2 we adopted a single geometric visualization of matrix games for two reasons: (1) all of the theory presented there could be interpreted in spatial terms with this model; (2) the Fundamental Theorem (Theorem 6) could be based directly on the geometrically obvious fact that the two convex sets which appeared could always be separated by a hyperplane. This is by no means the only model nor is this proof the shortest possible. In this appendix we will attempt a survey of the proofs of the Fundamental Theorem and, incidentally, present another geometric description. In the following chronological list we give references to all of the available proofs of this theorem (excluding proofs which include finite matrix games as a special case of infinite matrix games):

1. von Neumann, J., "Zur Theorie der Gesellshaftsspiele," *Math. Ann.*, **100** (1928), 295–320.

2. von Neumann, J., "Über ein ökonomisches Gleichungssystem und eine Verall-gemeinerung der Brouwerschen Fizpunktsatzes," *Ergebnisse eines Math. Kolloquiums*, **8** (1937), 73–83.

3. Ville, J., "Sur la théorie générale des jeux ou intervient l'habilité des joueurs traité du calcul des probabilités et de ses applications," by E. Borel and others, Paris (1938) Tome IV, Fasc. II, 105–113.

4. Kakutani, S., "A generalization of Brouwer's fixed point theorem," *Duke Math. Jour.*, **8** (1941), 457–459.

5. von Neumann, J., Morgenstern, O., *Theory of games and economic behaviour*, Princeton, 1944.

6. von Neumann, J., "A model of general economic equilibrium" (translation of [2]), *Rev. Econ. Studies*, **13** (1945–46), 1–9.

7. Loomis, L. H., "On a theorem of von Neumann," *Proc. Nat. Acad. Sci.*, **32** (1946), 213–215.

8. Dines, L. L., "On a theorem of von Neumann," *Proc. Nat. Acad. Sci.*, **33** (1947), 329–331.

9. Weyl, H., "Elementary proof of a minimax theorem due to von Neumann," in Annals of Math. Study No. 24.

10. Brown, G. W., von Neumann, J., "Solutions of games by differential equations," in Annals of Math. Study No. 24.

11. Gale, D., Kuhn, H. W., Tucker, A. W., "On symmetric games," in Annals of Math. Study No. 24.

12. Gale, D., "Convex polyhedral cones and linear inequalities," in *Activity Analysis of Production and Allocation*, edited by T. C. Koopmans, Wiley and Sons, New York, 1950.

13. Nash, J., "Equilibrium points in *n*-person games," *Proc. Nat. Acad. Sci.*, **36** (1950), 48–49.

14. Robinson, J., "An iterative method of solving a game," *Ann. of Math.*, **54** (1951), 296–301.

15. Nash, J., "Non-cooperative games," *Ann. of Math.*, **54** (1951), 286–295.

These proofs fall broadly into two classes, the first containing those proofs which are based on separation or support properties of convex sets, and the second composed of those proofs which use some notion of fixed point of a transformation or interactive procedure. We shall discuss these groups separately.

Proofs By Convex Sets

As was pointed out by Weyl in [9], Theorem 6 is a completely algebraic theorem and should be given an algebraic proof. Proofs [9] and [11] satisfy this restriction; the proof given in the text does not at two points. It is based ultimately on Theorem 2 where we use the topological fact that a continuous function on a closed bounded set in R_n assumes its minimum; this property is used again to pick v, the value of the game. These lapses are corrected by proving the algebraic equivalent of Theorem 2 and then reducing the general case to the case of symmetric matrix games.

THEOREM 2A. Given the vectors $U, U_1, \ldots, U_q$ in R_n, either $U = a_1 U_1 + \cdots + a_q U_q$ for non-negative $a_1, \ldots, a_q$ or there is an X such that $X \cdot U_\ell \geq 0$ for $\ell = 1, \ldots, q$ and $X \cdot U < 0$.

PROOF. (Clearly this is just the statement of Theorem 2 with $C = \{a_1 U_1 + \cdots + a_q U_q | a_1 \geq 0, \ldots, a_q \geq 0\}$.)

The theorem is proved by induction on n; it is obviously true for $n = 1$. We then note that the theorem is not changed by multiplication of $U, U_1, \ldots, U_q$ by *positive* constants. So we can assume (the theorem is trivially satisfied by $a_1 = \cdots = a_q = 0$ for $U = 0$)

$$U \cdot U = 1$$

and that the indices $\ell = 1, \ldots, q$ can be separated into three sets I, J, and K such that

$$U \cdot U_i = 1 \qquad \text{for } i \in I$$

$$U \cdot U_j = 0 \qquad \text{for } j \in J$$

$$U \cdot U_k = -1 \qquad \text{for } k \in K.$$

If I is empty, then $X = -U$ satisfies the second alternative of the theorem; otherwise, we project into the $(n - 1)$-dimensional space $H(U, 0)$ by considering the vectors

$$U_i - U \qquad \text{for } i \in I$$

$$U_j \qquad \text{for } j \in J$$

$$U_i + U_k \qquad \text{for } (i, k) \in I \times K.$$

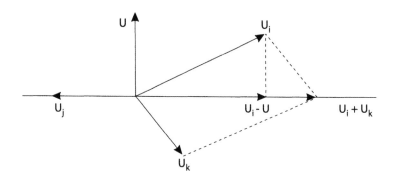

We then examine the solutions, T, of the inequalities

$$T \cdot (U_i - U) \geqq 0 \qquad \text{for } i \in I$$

$$T \cdot U_j \geqq 0 \qquad \text{for } j \in J$$

$$T \cdot (U_i + U_k) \geqq 0 \qquad \text{for } (i, k) \in I \times K.$$

There are two relevant cases.

CASE I. For each i there is a solution T_i such that $T_i \cdot (U_i - U) > 0$. Then $T = \sum_{i \in I} T_i$ is a solution and $T \cdot (U_i - U) > 0$ for all $i \in I$. Let $a = T \cdot U_{i_0} = \min_{i \in I} T \cdot U_i$ and set $X = T - aU$. Then

$$X \cdot U_i = T \cdot U_i - a \geq 0 \qquad\qquad \text{for } i \in I$$

$$X \cdot U_j = T \cdot U_j \geq 0 \qquad\qquad \text{for } j \in J$$

$$X \cdot U_k = X \cdot (U_{i_0} + U_k) = T \cdot (U_{i_0} + U_k) \geq 0 \quad \text{for } (i, k) \in I \times K.$$

and

$$X \cdot U = T \cdot U - a < 0,$$

hence the second alternative of the theorem is satisfied.

CASE 2. There is an i, say i_1, such that $T \cdot (U_{i_1} - U) = 0$ for all solutions T. Then we apply the induction hypothesis to $U_i - U_{i_1}$ and $U_i - U, U_j, U_i + U_k$. Since the second alternative does not hold in this case,

$$U - U_{i_1} = \sum_i a_i (U_i - U) + \sum_j b_j U_j + \sum_{(i,k)} c_{ik}(U_i + U_k)$$

where all a_i, b_j, and c_{ik} are non-negative. Hence

$$U = \left(U_{i_1} + \sum_i a_i U_i + \sum_j b_j U_j + \sum_{(i,k)} c_{ik}(U_i + U_k) \right) \Big/ \left(1 + \sum_i a_i \right)$$

satisfies the first alternative and the theorem is proved.

DEFINITION 13. A *symmetric* matrix game is a matrix game defined by a skew-symmetric matrix $A = (a_{ij})$, i.e., where $a_{ji} = -a_{ij}$.

Informally, in a symmetric matrix game the players have the same pure strategies when we take into account the (skew-symmetric) convention that a_{ij} measures the amount that P_2 pays to P_1. It is a trivial matter to show algebraically that Theorem 6 holds in general if and only if it holds for symmetric games (see [10] and [11] above).

THEOREM 6S. Every symmetric game has a solution.

PROOF. Apply Theorem 2A to the vectors

$$U \quad = (-1, -1, \ldots, -1)$$

$$U_\ell \quad = (a_{1\ell} \ldots, a_{m\ell}) \quad \text{for } \ell = 1, \ldots, m$$

$$U_{m+1} = (1, 0, \ldots, 0)$$

$$\cdots \quad \cdots\cdots$$

$$U_{2m} \quad = (0, 0, \ldots, 1).$$

If the first alternative holds then

$$-1 = \sum_{\ell=1}^{m} a_{i\ell}\, a_{\ell} + a_{m+i} \quad \text{for } i = 1, \ldots, m$$

where $a_1, \ldots, a_{2m}$ are non-negative. Multiplying the ith equation by a_i, adding, and taking account of the fact that $\sum_{i,\ell} a_i a_{i\ell}\, a_{\ell} = 0$ by the skew-symmetry of A, we have

$$-\sum_{i=1}^{m} a_i = \sum_{i=1}^{m} a_i\, a_{m+1} \,.$$

Hence $\sum_{i=1}^{m} a_i = 0$ or $U = a_{m+1} U_{m+1} + \cdots + a_{2m} U_{2m}$, which is a patent absurdity. Hence there is an $X = (x_1, \ldots, x_m)$ such that

$$x_1 a_{1\ell} + \cdots + x_m a_{m\ell} \geqq 0 \qquad \text{for } \ell = 1, \ldots, m$$

$$x_1 \geqq 0, \ldots, x_m \geqq 0$$

$$x_1 + \cdots + x_m > 0$$

If we set $\bar{X} = X \big/ (x_1 + \cdots + x_m)$ the theorem is proved.

It is possible to give a topological proof of Theorem 6 by means of another geometric realization. The model we propose has already been used implicitly for the discussion of the Skin Game in the heuristic preliminaries of Chapter 2. We consider a *game cylinder*, consisting of all $(m + 1)$-tuples $(x_1, \ldots, x_m, z)$ such that all $x_i \geq 0$, $x_1 + \cdots + x_m = 1$, and z is an arbitrary real number (see figure).

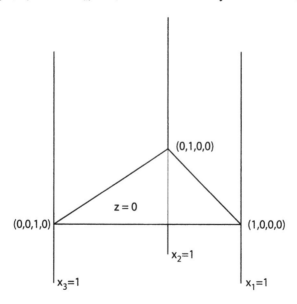

Then, if we are given the matrix game

$$
A = \begin{pmatrix} a_{11} & \cdots & a_{1n} \\ \vdots & & \vdots \\ a_{m1} & \cdots & a_{mn} \end{pmatrix}
$$

we can plot the expectation of P_1 when he uses his mixed strategy $X = (x_1, \ldots, x_m)$ against the jth pure strategy for P_2, $E(X, j) = a_{ij}x_1 + \cdots + a_{mj}x_m$ as the plane surface

(C_j) $\qquad\qquad\qquad\qquad z = a_{1j}x_1 + \cdots + a_{mj}x_m$

in the game cylinder. Thus, for the full Skin Game,

$$
\begin{pmatrix} 1 & -1 & -2 \\ -1 & 1 & 1 \\ 2 & -1 & 0 \end{pmatrix}
$$

we have three planes corresponding to the columns of the matrix:

(C_1) $\qquad\qquad\qquad\qquad z = x_1 - x_2 + 2x_3$
(C_2) $\qquad\qquad\qquad\qquad z = -x_1 + x_2 - x_3$
(C_3) $\qquad\qquad\qquad\qquad z = 2x_1 + x_2 .$

For each mixed strategy $X = (x_1, x_2, x_3)$, P_1 must count as his expectation the *minimum* z computed in this manner. If we graph this minimum surface in the game cylinder, it will appear as a polyhedral surface which is concave downward.

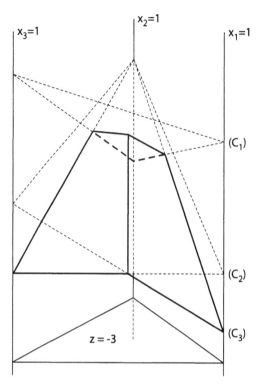

In this diagram we have graphed the three surfaces C_1, C_2, C_3 in light dotted lines, while the edges of the minimal surface are given in heavy outline. The surface $z = -3$ has been included to give a feeling for a horizontal cross-section of the cylinder.

The object of P_1 is to choose an $X = (x_1, x_2, x_3)$ which yields the highest z on the minimal surface, i.e., he wants to find the maximal z, such that for some X,

$$E(X, 1) = x_1 - x_2 + 2x_3 \geqq z$$

$$E(X, 2) = -x_1 + x_2 - x_3 \geqq z$$

$$E(X, 3) = -2x_1 + x_2 \qquad \geqq z$$

where at least one of the inequalities is an equation. Call v the maximal value of z and let $\bar{X}$ be an X which achieves it. (With our previous knowledge that the Fundamental Theorem holds, these clearly constitute half of a solution for the game.)

Where do the optimal strategies $\bar{Y}$ for P_2 appear in this model? It is easiest to answer this question by writing down the restrictions on such a strategy. They are

(1) $\qquad E(i, \bar{Y}) = a_{i1}\bar{y}_1 + \cdots + a_{in}\bar{y}_n \leqq v \quad$ for all i

(2) $\bar{y}_j \geq 0$ for all j and $\bar{y}_1 + \cdots + \bar{y}_n = 1$.

If we let $a_i = E(i, \bar{Y})$ for $i = 1, \ldots, m$ then the plane surfaces

$(\bar{C})$ $z = a_1 x_1 + \cdots + a_m x_m$

can be considered in a natural manner to be a *convex combination* of the surfaces
C_i). (The fact that a_i is the z-coordinate of the intersection of $(\bar{C})$ with the line
$x_i = 1$ and is also the ordinary convex combination of the corresponding coordinates a_{ij} for the surfaces C_j) strengthens this terminology.) Then conditions (1)
and (2) become: An optimal strategy for P_2 corresponds to a convex combination
of the surfaces (C_j) which lies wholly below $z = v$.

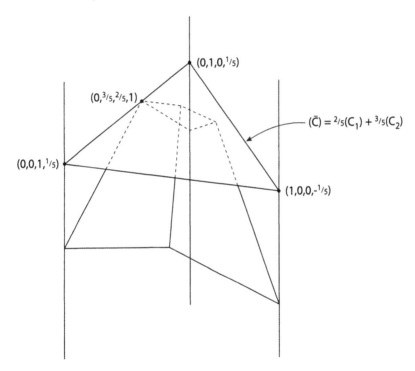

In our example, the highest point on the minimal surface is unique and equals
$(0, 3/5, 2/5, 1/5)$, corresponding to the unique optimal strategy $\bar{X} = (0, 3/5, 2/5)$
for P_1 and the value $v = 1/5$. To find the proper surface for P_2, remark that, since
one edge lies in the face of the cylinder, $x_2 + x_3 = 1$, passes through $(0, 3/5, 2/5,$
$1/5)$, and never has z coordinate greater than $1/5$, this edge has end points
$(0, 1, 0, 1/5)$ and $(0, 0, 1, 1/5)$. But such a surface $(\bar{C})$ can only be $2/5(C_1) +$
$3/5(C_2)$, corresponding to the unique optimal strategy $\bar{Y} = (2/5, 3/5, 0)$ for P_2.

How do we know, in general, that such a surface $(\bar{C})$ exists? Suppose that it did
not. Then, for all $Y = (y_1, \ldots, y_n)$ with all $y_j \geq 0$ and $y_1 + \cdots + y_n = 1$, there

would exist an i such that

$$a_{i1}y_1 + \cdots + a_{in}y_n > v.$$

But this says precisely that the convex hull of the columns of the game matrix (considered as vectors in R_m) does not intersect the corner S_v consisting of all $T = (t_1, \ldots, t_m)$ with $t_i \leqq v$ for $i = 1, \ldots, m$. Here we are presented with a separation problem of a much weaker sort than Theorem 5. We must only separate a closed bounded convex set (the convex hull of the columns) from a closed convex set (the corner) which it does not intersect. This is easily done by choosing a pair of points, U in the corner and V in the convex hull, which achieve the minimum distance between the sets, then applying the construction of Theorem 2:

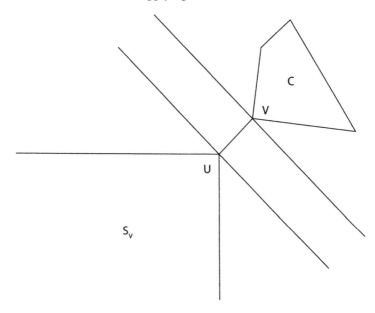

Then, by arguments exactly analogous to those used in Theorem 6, $X = V - U$ has non-negative components which may be normalized to sum to 1 and

$$X \cdot T > v$$

for all points T in the convex hull. Applying this to the columns, we have

$$x_1 a_{1j} + \cdots + x_m a_{mj} > v \quad \text{for } j = 1, \ldots, n$$

which contradicts the choice of v.

The technical difference between this proof and that given in the text is that between a proof by contradiction and a constructive proof. The constructive proof leaves us with a more satisfactory picture of the sets but needs the more refined Theorem 5. The proof by contradiction leaves us in some doubt as to where the optimal strategies lie. The proofs in [3], [5], and [8] are of this nature; the proof in [12] is closer to the proof given in Chapter 2.

Proofs By Fixed Points

Historically, the first successful proof of Theorem 6 was given by von Neumann with the use of an extension of the Brouwer fixed point theorem. This line of reasoning is found in proofs [1], [2], [4], [6], and [13]. In reference [6], Kakutani has given the most elegant statement of this point of view. However, Nash has recently succeeded [15] in giving a direct proof by means of the Brouwer theorem itself. His proof is so simple and beautiful that we include it here.

Given the matrix game $A = (a_{ij})$ we define

$$a_i(X, Y) = \text{Max } (0, E(i, Y) - E(X, Y)) \quad \text{for } i = 1, \dots, m$$

$$b_j(X, Y) = \text{Max } (0, E(X, Y) - E(X, j)) \quad \text{for } j = 1, \dots, n$$

and then give the continuous mapping $T : (X, Y) \longrightarrow (X', Y')$ of $M_1 \times M_2$ into $M_1 \times M_2$ by

$$x_i' = \frac{x_i + a_i}{1 + \sum_i a_1} \quad \text{for } i = 1, \dots, m$$

$$y_j' = \frac{y_j + b_j}{1 + \sum_j b_j} \quad \text{for } j = 1, \dots, n.$$

LEMMA. The mixed strategies $\bar{X}$ and $\bar{Y}$ are optimal if and only if $(\bar{X}, \bar{Y})$ is a fixed point under T.

PROOF. If $\bar{X}$ and $\bar{Y}$ are optimal then all of the a_i and b_j vanish. Hence $(\bar{X}, \bar{Y})$ is a fixed point.

Conversely, suppose $(\bar{X}, \bar{Y})$ is a fixed point. Since $E(\bar{X}, \bar{Y})$ is a convex combination of the $E(i, \bar{Y})$ for which $\bar{x}_i > 0$, we must have

$$E(\bar{X}, \bar{Y}) \geqq E(i, \bar{Y}) \quad \text{for some } i \text{ with } \bar{x}_i > 0.$$

Hence

$$a_i = 0 \quad \text{for some } i \text{ with } \bar{x}_i > 0$$

which implies that, since $\bar{x}_i$ is not decreased by T, $\sum_i a_i = 0$ or $a_i = 0$ for all i. Using the analogous argument for $\bar{Y}$ we have:

$$E(\bar{X}, j) \geqq E(\bar{X}, \bar{Y}) \geqq E(i, \bar{Y}) \text{ for all } i \text{ and } j,$$

and hence $\bar{X}$ and $\bar{Y}$ are optimal strategies.

To complete the proof, we remark that $M_1 \times M_2$ is a cell and hence the Brouwer fixed point theorem requires that T must have at least one fixed point $(\bar{X}, \bar{Y})$.

This proof is closely related to the differential equations proposed in [10]. If we restrict ourselves to symmetric matrix games and consider the transformation T defined above restricted to the "diagonal" of $M_1 \times M_2$ (which is a simplex) we have

$$x_i' = \frac{x_i + a_i}{1 + \sum_i a_i} \quad \text{for } i = 1, \ldots, m$$

where $a_i(X, X) = \text{Max}(0, E(i, X))$. Hence

$$x_i' - x_i = \left(a_i - \left(\sum_i a_i \right) x_i \right) \bigg/ \left(1 + \sum_i a_i \right)$$

or, dropping the normalization factor, we have

$$\frac{dx_i}{dt} = a_i - \left(\sum_i a_i \right) x_i \quad \text{for } i = 1, \ldots, m$$

which are the differential equations of Brown and von Neumann.

The method of fictitious play, proposed by Brown and shown to converge by J. Robinson [14] is also closely related to these transformations and provides another independent proof of Theorem 6.

Finally, we must mention the shortest known self-contained proof, due to Loomis [7], which is a straightforward proof by induction and uses no auxiliary machinery, thereby gaining in conciseness.

Chapter Three

Extensive Games

3.1 Some Preliminary Restrictions

By now the reader will have suspected that matrix games must play an important role in the theory of zero-sum two-person games and that they are much more inclusive than they may have appeared at first sight. Indeed, it may be asserted that *all finite zero-sum two-person games can be reduced to matrix games.* The critical reader may justly complain that this statement is meaningless since we have never defined what constitutes a game. Furthermore, although he understands vaguely what is meant by the word "game," his ideas on the subject are non-mathematical and give no clue as to how to verify the connection between these and the mathematical concept of a matrix game. The heart of the problem is close to this confusion; in fact, our assertion above was a non-mathematical assertion. To verify that it is true, we must provide an axiom system with its interpretation for finite games in general and show that this coincides, as an *interpreted system*, with the notion of a matrix game for zero-sum two-person games. We will be concerned only with the coincidence of certain aspects, indeed, exactly the strategic aspects which formed the basis of our definition of a solution. We have already solved this problem in the special instance of Simplified Poker; although the payoff matrix contained none of the verbal apparatus of the game given by its rules, we were convinced that it carried enough of the strategic possibilities to find a solution.

Therefore, our main problem in this chapter will be to give an axiom system with its interpretation which formalizes our intuitive notion of a game. The strict purist will not be satisfied with these axioms since they are framed in geometric language which could be axiomatized in turn. However, this is a mere quibble; we will make constant use of the intuitive insight given by the geometric model.

Throughout this chapter, passages that serve as motivation, interpretation, or heuristic discussion will be placed in square brackets [. . .]. This is done to emphasize the independence of the mathematical deductions of these sections.

3.2 The Axiom System

[In hunting for a geometrical scheme that will represent the combinatorial possibilities of an arbitrary game adequately, one might proceed by imagining the various possible plays as paths which begin at a single point 0 in an ordinary 2-dimensional Euclidean plane. Two distinct plays will branch at the first point at which they contain distinct choices; this branching point will be labeled with the name of the player, $P_1, \ldots, P_n$ making the choice or P_0 if it is made by a chance device. Thus, the four plays possible for Matching Pennies could be diagrammed as shown,

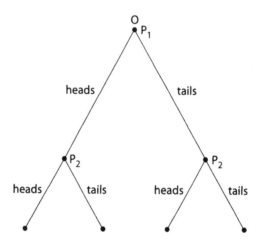

where we have split the one simultaneous choice into two successive choices. Since we are dealing with finite games at the moment, there will only be a finite number of possible choices at each vertex in our figure and each play (i.e., path) will contain only a finite number of edges joining vertices [1]. We can also draw our figure so that there is no ambiguity in how we reach a given position from 0 by requiring that there are no closed paths in the figure. Such figures (connected graphs with no cycles) appear often in modern mathematics and are called, quite properly, *trees*.]

DEFINITION 14. A *game tree K* is a finite tree with a distinguished vertex 0 which is embedded in an oriented plane.

[As we have seen, a game tree is a natural geometric model for the essential character of a game as a successive presentation of alternatives. The distinguished vertex represents the first occasion of a choice; the embedding in an oriented plane is a convenient device for numbering alternatives. Thus, by agreeing that we will number in the positive sense of the orientation starting at a vertex X with the first alternative which is not on the path from 0 to X we can draw

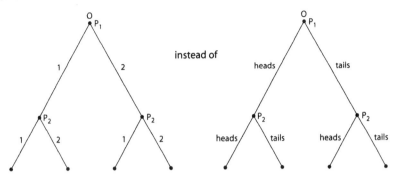

as the diagram for Matching Pennies. This indexing is important for two functions, for recording the course of a play and later for making the correspondence between alternatives which appear the same to a player on the basis of the information that is available to him.

Before proceeding to the definition of a game, it is necessary to present some general technical terms associated with a game tree; it is important to remark that, although these terms are taken from common parlance, their meaning is given by the definitions.]

TERMINOLOGY. The vertices $X, Y, Z, \ldots$ of a game tree K are called *positions*. If K is cut into at least two non-void components by removing the position X, we shall call X a *move*; otherwise X is called a *play* and denoted by the generic symbol W. The edges e incident at a move X which are not connected to 0 when we delete X are called the *alternatives* at X. If there are j alternatives at a move X then they are indexed by the integers $1, \ldots, j$ circling X in the positive sense of the orientation. At the position 0, the first alternative may be assigned arbitrarily. If one circles a move $X \neq 0$ in the positive sense, the first alternative follows the unique edge incident to X which is not an alternative. The function that indexes the alternatives in K will be denoted by ν; thus $\nu(e)$ is the index of the alternative e. The set of moves is partitioned [2] into sets $A_1, \ldots, A_j, \ldots$, where A_j contains those moves with j alternatives; this partition will be called the *alternative partition* and denoted by $\mathcal{A}$.

[The specifications which are lacking in this model are quite clear. We must add (1) the assignment of moves to players or chance, (2) the assignment of the chance probabilities, (3) the assignment of the payoff at the ends of the plays, and (4) the specification of the state of information of a player when he is called upon to make a choice. The first three requirements are easy to meet in a straightforward manner; the fourth is more difficult. To see how the problem of information is solved, we remark that the rules of the game must specify the information that is given to a player on the occasion of a choice, i.e., at a move X. Let us denote this information by a function $\mathcal{I}(X)$ which depends on the move X. (An argument can be made that, since the function $\mathcal{I}$ is defined by the verbal rules on the finite set of moves, we can assume $\mathcal{I}(X)$ is an integer; however, this is irrelevant to our heuristic discussion. We need only distinguish when $\mathcal{I}(X) = \mathcal{I}(Y)$.) If the function $\mathcal{I}$ assumes the same value on X and Y, i.e., if a player is given the same information when the play has progressed to either X or Y we shall say that X *and Y belong to the same information set U*. It is clear that this defines a partition $\mathcal{U}$ of the moves of the game into information sets and, since a player only uses the information to deduce the actual move that is presented him, we will assume that $\mathcal{I}(X)$ is the information set U which contains X. Since all of the players are in possession of the rules of the game, if two moves have a different number of alternatives or if they are assigned to different players then they cannot be in the same information set. We will also assume that two moves in the same information set never occur in the same play [3].]

DEFINITION 15. An *n-person game*, Γ, *in extensive form* is a game tree K with the following specifications:

(I) A partition $\mathcal{P}$ of the moves into $n + 1$ indexed sets P_0, $P_1, \ldots, P_n$ which will be called the *player partition*. The moves in P_0 will be called *chance moves*; the moves in P_i, for $i = 1, \ldots, n$, will be called *personal moves of player i*.

(II) For each chance move with j alternatives, a probability distribution on the alternatives $1, \ldots, j$, which assigns *positive* probability to each alternative.

(III) For each play W, an n-tuple of real numbers, $h(W) = (h_1(W), \ldots, h_n(W))$. The vector valued function h will be called the *payoff function*.

(IV) A partition $\mathcal{U}$ of the moves into sets U which is a refinement of the player and alternative partitions (that is, each U is contained in some $P_i \cap A_j$) and such that no U contains two moves on the same play.

[How is this formal scheme to be interpreted? That is, how does one play an n-person game in extensive form? To personalize the interpretation, one may imagine a number of people called *agents* isolated from each other and each in possession of the rules of the game. There is one agent for each information set and they are grouped into players in a natural manner, an agent belonging to the ith player if his information set lies in P_i. This apparent plethora of agents is occasioned by the possibly complicated state of information of our players who may be forced by the rules to forget information that they knew earlier in a play.

A play begins at the vertex 0. Suppose that it has progressed to the move X. If X is a personal move with j alternatives then the agent whose information set U contains X chooses a positive integer not greater than j, knowing only that he is choosing an alternative with this index at one of the moves in U. If X is a chance move, then an alternative is chosen by a chance device in accordance with the probabilities assigned to the alternatives at X by (II). In this manner, a path with initial point 0 is constructed. It is unicursal and, since K is finite, leads to a unique play W. At this point, P_i is paid the amount $h_i(W)$ for $i = 1, \ldots, n$. The case in which K reduces to the vertex 0 is not excluded; then Γ is a no move game, no one does anything, and the payoff is $h(0)$.]

EXAMPLE. The game of Matching Pennies is now defined as in the figure,

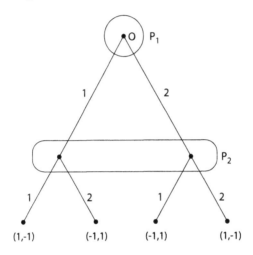

where we have labeled the alternatives with their indices and circled the moves in the same information set.

EXAMPLE. The extensive form of an $m \times n$ matrix game is as shown,

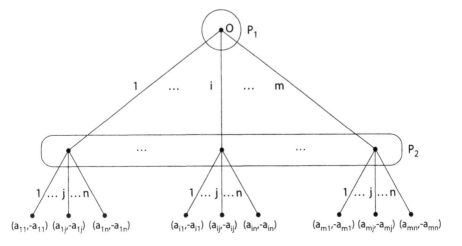

where the fact that all of the moves of P_2 lie in the same information set reflects his ignorance of the choice of P_1.

EXERCISE 9. Give the extensive form of Simplified Poker.

3.3 Pure and Mixed Strategies

[As we have interpreted a pure strategy, say in Simplified Poker, it is a plan formulated before a play by a *strategist* for a player. He then communicates his choices to the agents of this player. One may do this without violating the rules of the game regarding the information possessed by the agents by imagining that the strategist writes a book with one page for each of his information sets U. If each move in U has j alternatives, the corresponding page will contain a positive integer not greater than j; this page is given to the agent who is acting on that information set. If he is called upon during a play he is to choose the alternative indexed by that integer. These remarks motivate the following definitions.]

> **DEFINITION 16.** Let $\mathcal{U}_i = \{U | U \in P_i\}$.
> A *pure strategy for* P_i, for $i = 1, \ldots, n$, is a function π_i mapping $\mathcal{U}_i$ into the positive integers such that $U \subset A_j$ implies $\pi_i(U) \leqq j$. We will say that π_i *chooses* the alternative e at $X \in U$ if $\pi_i(U) = \nu(e)$.
> An n-tuple $\pi = (\pi_1, \ldots, \pi_n)$ of pure strategies defines a probability distribution on the alternatives in each of the information sets U as follows:
> If e is an alternative at $X \in U \subset P_0$, then $P_e(\pi)$ is the probability assigned to $\nu(e)$ by (II).
> If e is an alternative at $X \in U \subset P_i$, for $i = 1, \ldots, n$, then $p_e(\pi) = 1$ if $\pi_i(U) = \nu(e)$ and $p_e(\pi) = 0$ otherwise.
> This, in turn, defines the *probability that a position X will occur* by:
> $p_X(\pi) = \prod_{e \subset OX} p_e(\pi)$ where OX denotes the path from 0 to X. Then the *expected payoff,* $E_i(\pi)$, *to* P_i is defined by
>
> (1) $E_i(\pi) = \sum_W p_W(\pi) h_i(W)$ for $i = 1, \ldots, n$.

[The interpretation of each of these notions is indicated by their names and the preceding discussion. Since the tree is finite, there are only a finite number of pure strategies and hence expression (1) is a n-dimensional matrix form for the game. This may become clearer if we write down the case for $n = 2$. There

$$E_1(\pi_1, \pi_2) = \sum_W p_W(\pi) h_1(W)$$

$$E_2(\pi_1, \pi_2) = \sum_W p_W(\pi) h_2(W)$$

where π_1 and π_2 range over finite sets, π_1 corresponding to the rows of the matrix and π_2 to the columns. It is only when $E_1(\pi) + E_2(\pi) = 0$, i.e., in the zero-sum case, that we can use a single matrix in a two-person game.

As we soon found for matrix games, it is not rational to play the same pure strategy on all plays; it is necessary to randomize in some manner to protect against being "found out" by your opponent. Thus a mixed stragegy will attach a probability to each book (= pure strategy) and before the play begins the strategist will choose the book he will use by means of a chance device in accordance with these probabilities. Thus, the player will still follow a fixed plan throughout the course of a play but this plan will be chosen by a chance device before the play begins.]

DEFINITION 17. A *mixed strategy for* P_i, for $i = 1, \ldots, n$, is a probability distribution μ_i on the pure strategies for P_i, which assigns the probability q_{π_i} to π_i.

An n-tuple $\mu = (\mu_1, \ldots, \mu_n)$ of mixed stratgegies defines the *probability that a position X will occur* by $p_X(\mu) = \sum_\pi q_{\pi_1} \cdots q_{\pi_n} p_X(\pi)$. Then the *expected payoff*, $E_i(\mu)$, to P_i is defined by:

$$(2) \qquad E_i(\mu) = \sum_W p_W(\mu)\, h_i(W) \quad \text{for } i = 1, \ldots, n.$$

[Now we are in a position to verify the assertion made in the first section, namely, that all finite zero-sum two-person games can be reduced to matrix games. This is done by constructing the following table of interpretational correspondences:

	Matrix Games	Extensive Games
Pure strategies	π_1, π_2	π_1 , π_2
Plays	$\pi = (\pi_1, \pi_2)$	W
Payoff to P_1 on a play	$\sum_W p_W(\pi) h_1(W)$	$h_1(W)$
Mixed strategies	μ_1, μ_2	μ_1, μ_2
Payoff to P_1 on mixed strategies	$\sum_\pi q_{\pi_1} q_{\pi_2} \left(\sum_W p_W(\pi) h_1(W) \right) = \sum_W \left(\sum_\pi q_{\pi_1} q_{\pi_2} p_W(\pi) \right) h_1(W)$	

Thus, although the notion of a play is quite distinct in the two formulations, we have exact correspondence of the strategic concepts and of the expected payoffs.]

3.4 Games with Perfect Information

DEFINITION 18. A game with *perfect information* is a game which the information partition consists of one element sets.

[It is clear that perfect information means that a player is informed at every move of the exact sequence of choices preceding that move. Examples of games with perfect information are chess, checkers, backgammon, and tic-tac-toe. They are distinguished by the interesting fact that one need not randomize when playing them; for example, there is a best pure strategy in chess which cannot be improved by mixing it with other strategies. We will prove this result only for zero-sum two-person games although it holds in a more general situation when properly interpreted [4].]

THEOREM 9. Every zero-sum two-person game with perfect information has a solution in pure strategies.

PROOF. To insure the clarity of the statement, remark that the payoff of a zero-sum two-person game satisfies:

$$E_1(\pi_1, \pi_2) + E_2(\pi_1, \pi_2) = 0 \quad \text{for all } \pi_1 \text{ and } \pi_2$$

and, by a solution in pure strategies we mean pure strategies $\bar{\pi}_1$, $\bar{\pi}_2$ such that:

$$E_1(\bar{\pi}_1, \pi_2) \geqq E_1(\bar{\pi}_1, \bar{\pi}_2) \quad \text{for all } \pi_2$$

$$E_2(\pi_1, \bar{\pi}_2) \geqq E_2(\bar{\pi}_1, \bar{\pi}_2) \quad \text{for all } \pi_1 .$$

We will prove this theorem by induction on the length of the game Γ, that is, on the maximum number of moves in a play of Γ. The induction is easily begun, since the theorem is trivially true for a game of length zero. To achieve the induction step, one subgame is defined for each alternative at the move O in Γ. Assume that there are j alternatives at O and that these lead to the moves $O_1, \ldots, O_j$. The subgame Γ_ν, $\nu = 1, \ldots, j$ is then defined as follows:

The game tree K_ν is the component of K that contains O if we retain only the νth alternative at O, i.e., it consists of O_ν and all of the moves that follow O_ν. As such it is imbedded in the same oriented plane as K and has O_ν as distinguished vertex. The player and information partitions of the moves of K_ν are the respective partitions of the moves of K restricted to K_ν. For each chance move in K_ν the chance probabilities are those assigned to the alternatives at that move by (II). The payoff function h_ν for K_ν is the payoff function h restricted to the plays of K_ν.

The main burden of the proof rests on an analysis of the relations between the pure strategies $\pi_{v/i}$ and the payoffs $h_{v/i}$ of these games and those of Γ. This analysis falls naturally into two cases:

CASE 1. The first move O of Γ is a personal move, say of P_1. Then the induction hypothesis says that, for $v = 1, \ldots, j$, there exist pure strategies $\bar{\pi}_{v/1}$ and $\bar{\pi}_{v/2}$ for Γ_v such that

$$E_{v/1}(\bar{\pi}_{v/1}, \pi_{v/2}) \geqq E_{v/1}(\bar{\pi}_{v_1}, \bar{\pi}_{v_2}), \quad \text{for all } \pi_{v/2}$$

$$E_{v/2}(\pi_{v/1}, \bar{\pi}_{v/2}) \geqq E_{v/2}(\bar{\pi}_{v/1}, \bar{\pi}_{v/2}) \quad \text{for all } \pi_{v/1}.$$

Then we construct an optimal pure strategy for P_1 by choosing $\bar{v}$ so as to maximize $E_{v/1}(\bar{\pi}_{v/1}, \bar{\pi}_{v/2})$ and playing as is dictated by the $\bar{\pi}_{v/1}$ in the subgames Γ_v. For P_2, the choices in the Γ_v are given by the $\bar{\pi}_{v/2}$. Then, if $\bar{\pi}_1$ is a pure strategy that chooses $\bar{v}$ at O and coincides with $\bar{\pi}_{v/1}$ in Γ_v and $\bar{\pi}_2$ is a pure strategy that coincides with $\pi_{v/2}$ in Γ_v:

$$E_1(\bar{\pi}_1, \pi_2) = E_{\bar{v}/1}(\bar{\pi}_{\bar{v}/1}, \pi_{\bar{v}/2}) \quad \geqq \quad E_{\bar{v}/1}(\bar{\pi}_{\bar{v}/1}, \bar{\pi}_{\bar{v}/2}) = E_1(\bar{\pi}_1, \bar{\pi}_2)$$

$$E_2(\pi_1, \bar{\pi}_2) = E_{v/2}(\pi_{v/1}, \bar{\pi}_{v/2}) \quad \geqq \quad E_{v/2}(\bar{\pi}_{v/1}, \bar{\pi}_{v/2}) \geqq E_{\bar{v}/1}(\bar{\pi}_{\bar{v}/1}, \bar{\pi}_{\bar{v}/2})$$

$$= E_2(\bar{\pi}_1, \bar{\pi}_2)$$

CASE 2. The first move O of Γ is a chance move, with j alternatives, $v = 1, \ldots j$; the vth alternative is to assign probability p_v by (II). Then we construct optimal strategies by having P_1 play in accordance with $\bar{\pi}_{v/1}$ in Γ_v and P_2 play in accordance with $\bar{\pi}_{v/2}$ where these are chosen with the use of the induction hypothesis

$$E_1(\bar{\pi}_1, \pi_2) = \sum_v p_v E_{v/1}(\bar{\pi}_{v/1}, \pi_{v/2}) \geqq \sum_v p_v E_{v/1}(\bar{\pi}_{v/1}, \bar{\pi}_{v/2})$$

$$= E_1(\bar{\pi}_1, \bar{\pi}_2)$$

$$E_2(\pi_1, \bar{\pi}_2) = \sum_v p_v E_{v/2}(\pi_{v/1}, \bar{\pi}_{v/2}) \geqq \sum_v p_v E_{v/2}(\bar{\pi}_{v/1}, \bar{\pi}_{v/2})$$

$$= E_2(\bar{\pi}_1, \bar{\pi}_2).$$

This completes the proof of the theorem.

3.5 A Reduction of the Game Matrix

[If we apply Definition 16 directly to Simplified Poker, we find that, since P_1 has six information sets with two alternatives each, there are 2^6 pure strategies for P_1.

However, when we "normalized" this game (see page 38), we found that P_1 had but 27 pure strategies. How is this discrepancy to be explained?

The answer is quite simple; if Definition 16 is followed strictly, P_1 will make a plan for the second round even if he decides to bet in the first round. But if he bets in the first round the play always ends after P_2's choice and none of the moves in the second round can possibly occur. Therefore his plans for the second round are completely irrelevant. Of course, it does no harm to formulate such strategies; it only produces duplicated rows or columns in the matrix since these decisions do not alter the expectation of the player involved. A skeptical reader may object: "Don't the irrelevant choices become operative when you mix pure strategies and thus influence the expectation?" A negative answer is readily obtained by recalling that, when playing a mixed strategy, a single pure strategy is chosen before each play of the game and is used consistently throughout that play. Hence the irrelevant choice of one pure strategy can never change another pure strategy through mixing. However, in Section 7 we shall discuss another method of randomizing in which the irrelevant choices may have an effect and so it is essential that they be removed from our notion of pure strategy now. Incidentally, this will reduce the matrix by striking out repeated rows and columns [5].]

DEFINITION 19. The set of positions which are *possible when playing* π_i, Poss (π_i), consists of all X such that $p_X(\pi) > 0$ for some $\pi = (\pi_1, \ldots, \pi_i, \ldots, \pi_n)$. The set of all information sets which are *relevant when playing* π_i, Rel (π_i), consists of all $U \subset P_i$ such that $U \cap$ Poss (π_i) is non-void.

Two pure strategies π_i' and π_i'' for P_i are said to be *equivalent*, $\pi_i' \equiv \pi_i''$, if Rel $(\pi_i') =$ Rel (π_i'') and $\pi_i'(U) = \pi_i''(U)$ for all $U \in$ Rel (π_i').

It is clear that the relation of equivalence is symmetric, reflexive, and transitive and therefore defines a partition of the pure strategies into equivalence classes [6]. The next theorem says, in effect, that equivalent pure strategies give the same expected payoff against all counter-strategies for the opposing players; that is, they correspond to rows or columns with the same entries in the game matrix. In the proof of this theorem and in the subsequent text it is convenient to let π/π_i' denote $(\pi_1, \ldots, \pi_{i-1}, \pi_i', \pi_{i+1}, \ldots, \pi_n)$ where $\pi = (\pi_1, \ldots, \pi_i, \ldots, \pi_n)$ is any n-tuple of pure strategies.

THEOREM 10. Two pure strategies π_i' and π_i'' for P_i are equivalent if and only if

$$p_W \left(\pi/\pi_i' \right) = p_W \left(\pi/\pi_i'' \right)$$

for all plays W and all $\pi = (\pi_1, \ldots, \pi_n)$.

PROOF. To show necessity, suppose that $p_W(\pi/\pi_i') > 0$. Then $p_X(\pi/\pi_i') > 0$ for all positions X in the play W and hence all such X are possible when playing π_i'. Therefore all information sets U for P_i which contain moves X for P_i in W are relevant when playing π_i', which implies that

$$U \in \text{Rel}(\pi_i'') \quad \text{and} \quad \pi_i'(U) = \pi_i''(U)$$

for such U. Stated verbally, π_i' and π_i'' make the same choices at all moves for P_i in the play W. Hence $p_W(\pi/\pi_i') > 0$. If we interchange π_i' and π_i'' we find that $p_W(\pi/\pi_i) > 0$ implies $p_W(\pi/\pi_i') > 0$ or $p_W(\pi/\pi_i') = 0$ implies $p_W(\pi/\pi_i) = 0$ and the necessity is proved.

To prove sufficiency, suppose $U \in \text{Rel}(\pi_i')$. Then there are an $X \in U$ and a π such that $p_X(\pi/\pi_i') > 0$. Complete OX to a play W by following the choices of π/π_i' using arbitrary alternatives at chance moves. Then $p_W(\pi/\pi_i') > 0$. Therefore $p_W(\pi/\pi_i') > 0$ which implies that $\pi_i(U) = \pi_i''(U)$ and $p_X(\pi/\pi_i'') > 0$ or $U \in \text{Rel}(\pi_i'')$.

COROLLARY. If $\pi_j' \equiv \pi_j''$ then we have $E_i(\pi/\pi_j') = E_i(\pi/\pi_j'')$ for all π and $i = 1, \ldots, n$.

PROOF. This is an immediate consequence of Theorem 10 and the definition of the expected payoffs E_i.

It should be noted that a converse to this corollary is also true; namely, given $\pi_j' \not\equiv \pi_j''$, we can always construct a payoff function h such that

$$E_i\left(\pi/\pi_j'\right) \neq E_i\left(\pi/\pi_j''\right) \quad \text{for some } i \text{ and some } \pi.$$

Since $\pi_j' \not\equiv \pi_j''$, Theorem 10 implies that there exist a W_0 and a π such that $p_{W_0}(\pi/\pi_j') \neq p_{W_0}(\pi/\pi_j'')$. We can assume

$$p_{W_0}\left(\pi/\pi_j'\right) > p_{W_0}\left(\pi/\pi_j''\right) \geqq 0.$$

Now define

$$h_i(W) = 0 \quad \text{for } W \neq W_0 \text{ and } i = 1, \ldots, n$$

$$h_i(W_0) = 0 \quad \text{for } i \neq j$$

$$h_j(W_0) = 1.$$

Then

$$E_j\left(\pi/\pi_j'\right) = p_{W_0}\left(\pi/\pi_j'\right) > p_{W_0}\left(\pi/\pi_j''\right) = E_j\left(\pi/\pi_j''\right).$$

This proves our contention.

3.6 An Instructive Example

[The following game imitates, in the most rudimentary manner, a situation that arises frequently; it will give us a valuable insight into a method of treating a wide class of games in extensive form.

EXAMPLE. The cards in an ordinary deck are given numerical values, $1, 2, \ldots, 13$ (disregarding suit, these may be Ace through King). Two cards are dealt at random; one is concealed and the other is shown to the single player P_1. He then must decide whether to drop out or bet. If he drops out, he must pay two units for playing; if he bets, he wins the numerical difference between his card and the concealed card (paying if this difference is negative).

This is another, qualitatively different situation in which a direct application of Definition 16 would be foolhardy. Once we remark that P_1 has thirteen information sets, each with two alternatives, and thus has $2^{13} = 8192$ pure strategies it becomes clear that their mere enumeration is quite a tedious task. However, let us imagine that we are kibbitzing the play of this game. As we observe a large number of plays, a striking fact emerges. Since we only see one choice in each play and therefore cannot tell what choice P_1 would have made had he been dealt a different card, the deduction of his pure strategy on that play is impossible. Hence, *we are not able to observe the mixed strategy that P_1 is using.* Instead, we observe his behavior independently in the thirteen possible situations that confront him. For example, we can make an estimate of the probabilities that he will drop out or bet when he is dealt a 6.

Let us examine this situation more closely. A pure strategy π_1 for P_1 is a vector $(a_1, \ldots, a_{13})$ with thirteen components (the pages of his strategy book) where

$$a_k = \begin{cases} 0 \\ 1 \end{cases} \text{ means } \begin{cases} \text{drop out} \\ \text{bet} \end{cases} \text{ for } k = 1, \ldots, 13.$$

Hence, a mixed strategy μ_1 for P_1 is an assignment of a probability $q(a_1, \ldots, a_{13})$ to each of his pure strategies and hence his mixed strategies form a (8191)-dimensional simplex. On the other hand, the behavior that we observe can be described by a vector $\beta = (b_1, \ldots, b_{13})$ where

b_k = probability that P_1 drops out when dealt card k.

Hence b_k satisfies $0 \leq b_k \leq 1$ and the "behaviors" form a 13-dimensional "cube."

Suppose P_1 plays his mixed strategy μ; which "behavior" β do we observe? It is clear that

$$b_k = \sum_{a_k = 0} q(a_1, \ldots, a_{13}) \quad \text{for } k = 1, \ldots, 13$$

defines the observed β corresponding to the mixed strategy with probabilities $q(a_1, \ldots, a_{13})$. Note that many mixed strategies dictate the same behavior! Stated mathematically, the mapping $\mu \longrightarrow \beta(\mu)$ is a many-one mapping, indeed, reduces the dimension from 8191 to 13.

Now let us use these considerations to solve our game. Instead of using a mixed strategy we will use a *behavior strategy*; that is, we will choose the probabilities b_k with which we will drop out if dealt card k. Since the deals are independent random events we can choose b_k independently for each k. To do this, it is clear that we need only compare the cost of dropping out with the expectation if we bet. These expectations are tabulated in the following table:

Card $k =$	1	2	3	4	5	6	7
Expectation	$\mp\dfrac{312}{51}$	$\mp\dfrac{260}{51}$	$\mp\dfrac{208}{51}$	$\mp\dfrac{156}{51}$	$\mp\dfrac{104}{51}$	$\mp\dfrac{52}{51}$	0
Card $k =$	13	12	11	10	9	8	7

where the expectations are negative in the upper row and positive in the lower row. It is only rational for P_1 to bet when, by so doing, his expectation is greater than -2. Hence his single optimal behavior strategy is

$$\beta = (0, 0, 0, 0, 0, 1, 1, 1, 1, 1, 1, 1, 1)$$

that is, he should bet when dealt a 6 or higher. His expectation is then

$$\frac{1}{13}\left(-2 - 2 - 2 - 2 - 2 - \frac{52}{51} + \frac{52}{51} + \frac{104}{51} + \cdots + \frac{312}{51}\right) = \frac{530}{13.51} \approx 0.8.$$

Therefore, this is quite a desirable game to play; in order to make a reasonable profit, the "house" should charge 1 unit for the rent of the cards.

However, it is not the solution of this rather special example that interests us; it is rather the phenomenon of a radical reduction in the dimension through the use of "behavior strategies" that gives this example importance. Before we make a formal definition it is instructive to interpret them through the use of the notion of a strategy book. In these terms, a behavior strategy book has just as many pages as a pure strategy book (one for each information set) and lists on each page not a single choice but a probability distribution on the alternatives available in the situation corresponding to that page. Thus, in cribbage, if the pegs on the board are at a given position and the cards have been dealt, a player would look up that situation in his behavior strategy book and find the ten probabilities with which he should discard the ten possible pairs of cards from his five card hand.

One final remark must be made relevant to the mapping of mixed strategies onto behavior strategies and the reduction of Section 5. The mapping is defined in the following manner: A mixed strategy assigns a probability to each pure strategy book. To find the behavior strategy probability assigned to a fixed alternative at a fixed page (= information set), examine that page in all of the pure strategy books and add up the probabilities assigned to books which choose the given alternative.

This procedure would be satisfactory were it not for irrelevant information sets; if we take them into consideration they may cause us to increase the probability of alternatives which we do not actually choose in play. Therefore, in accordance with Section 5, we will consider an equivalence class of pure strategies instead of a pure strategy. By Definition 19, these may be interpreted as strategy books with blank pages for the irrelevant information sets. Defining a mixed strategy to be a probability assignment to these new strategy books, we can follow our definition of the mapping given above except that we must divide the sum of the probabilities by the sum of the probabilities assigned to books in which the given page is not blank. If this sum is zero, we will say that that page is irrelevant for the given mixed strategy and leave it blank in the corresponding behavior strategy book. Obviously, this is satisfactory since the situation to which it corresponds must always have been rendered impossible by a previous choice.]

3.7 Behavior Strategies and Perfect Recall

Throughout this section, pure strategy will mean an equivalence class of pure strategies following Defintion 19; the definition of mixed strategy is to be changed accordingly. We do not change our notation.

DEFINITION 20. Let the mixed strategy μ_i assign the probability q_{π_i} to each pure strategy π_i for P_i and consider the information set $U \subset P_i \cap A_j$. Let $S = \{\pi_i | U \text{ is relevant for } \pi_i\}$. Then S is partitioned into j sets $S_\nu = \{\pi_i | \pi_i(U) = \nu\}$, $\nu = 1, \ldots, j$. If

$$\sum_{\pi_i \in S} q_{\pi_i} \neq 0$$

then U is said to be *relevant when playing μ_i* and

$$b_\nu = \sum_{\pi_i \in S_\nu} q_{\pi_i} \bigg/ \sum_{\pi_i \in S} q_{\pi_i} \text{ for } \nu = 1, \ldots, j$$

defines a distribution on $1, \ldots, j$. The aggregate of all such assignments on information sets which are relevant for μ_i is called the *behavior strategy associated with μ_i* and is denoted by $\beta_i(\mu_i)$ or simply by β_i.

Any n-tuple $\beta = (\beta_1, \ldots, \beta_n)$ of behavior strategies for the players $P_1, \ldots, P_n$ defines $p_X(\beta)$, *the probability that a position X will occur*, as follows:

If e is an alternative at a move for P_i, $i = 1, \ldots, n$, in the relevant information set U, let $p_e(\beta) = b_{\nu(e)}$.

If e is an alternative at a chance move, let $p_e(\beta)$ be the probability assigned to $\nu(e)$ by (II).

Otherwise, let $p_e(\beta) = 0$.

Then

$$p_X(\beta) = \prod_{e \subset OX} p_e(\beta).$$

The *expected payoff*, $E(\beta)$, is defined by

$$E_i(\beta) = \sum_W p_W(\beta) h_i(W), \quad i = 1, \ldots, n.$$

Since we have derived behavior strategies from equivalence classes, it can be shown that two behavior strategies are *identical* if and only if they yield the same $p_W(\beta)$ for all plays W and all counter-strategies for the opposing players. However, this will not be needed here.

[It might be thought that we can solve all games through the use of behavior strategies. The following example counters this assertion and indicates the necessary restriction.]

EXAMPLE. Solve the zero-sum two-person game shown using mixed strategies.

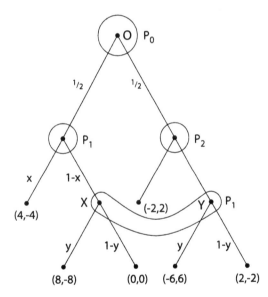

How much can P_1 assure himself using behavior strategies?

The matrix of this game is easily calculated to be

$$
\begin{array}{cc}
 & (1) \quad (2) \\
\begin{array}{c}
(1,1) \\
(1,2) \\
(2,1) \\
(2,2)
\end{array}
&
\left(
\begin{array}{cc}
1 & -1 \\
1 & 3 \\
3 & 1 \\
-1 & 1
\end{array}
\right)
\end{array}
$$

where we have labeled the rows with the pure strategies of P_1 and P_2. The mixed strategies $\bar{\mu}_1 = (0, 1/2, 1/2, 0)$ and $\bar{\mu}_2 = (1/2, 1/2)$ are clearly the unique optimal mixed strategies and yield the value $v = 2$. If we denote by (x, y) the behavior strategy indicated on the tree for P_1:

$$E_1((x, y), (1)) = 2x + 4y - 4xy - 1$$

$$E_1((x, y), (2)) = 2x \qquad - 4xy + 1.$$

Hence

$$E_1((x, y), (1)) \leqq E_1((x, y), (2)) \quad \text{when } 0 \leqq y \leqq \frac{1}{2},$$

and

$$E_1((x, y), (2)) \leqq E_1((x, y), (1)) \quad \text{when } \frac{1}{2} \leqq y \leqq 1.$$

Taking a safe viewpoint, we must find

$$\text{Max } (2x + 4y - 4xy - 1) \quad \text{for } 0 \leqq x \leqq 1 \text{ and } 0 \leqq y \leqq \frac{1}{2},$$

and

$$\text{Max } (2x - 4xy + 1) \quad \text{for } 0 \leqq x \leqq 1 \text{ and } \frac{1}{2} \leqq y \leqq 1.$$

By straightforward calculation,

$$\underset{\substack{0 \leqq x \leqq 1 \\ 0 \leqq y \leqq 1/2}}{\text{Max}} (2x + 4y - 4xy - 1) = \underset{\substack{0 \leqq x \leqq 1 \\ 0 \leqq y \leqq 1/2}}{\text{Max}} (2x(1 - 2y) + 4y - 1)$$

$$= \underset{0 \leqq y \leqq 1/2}{\text{Max}} (1 - 2y + 4y - 1) = 1$$

$$\underset{\substack{0 \leqq x \leqq 1 \\ 1/2 \leqq y \leqq 1}}{\text{Max}} (2x - 4xy + 1) = \underset{\substack{0 \leqq x \leqq 1 \\ 1/2 \leqq y \leqq 1}}{\text{Max}} (2x(1 - 2y) + 1) = 1.$$

Therefore, the most P_1 can assure himself using behavior strategies is 1 and this is obtained by playing any x and $y = 1/2$.

[The precise reason why mixed strategies are superior to behavior strategies in this example is that, through their use, P_1 can distinguish the plays in which he has intentionally carried the game to a third choice and those in which it has proceeded to a third choice through the action of P_2. That is, when he has played the pure strategies $(1, 1)$ or $(1, 2)$, he knows that the move X is not possible and hence, if he is presented with a third choice, he must be at Y. Stated in verbal terms, he has forgotten what he knew immediately after the chance move, but the use of pure strategies enables him to communicate some of this forgotten information.]

DEFINITION 21. An n-person game is said to have *perfect recall* if

$$X \in U \in \text{Rel}(\pi_i) \implies X \in \text{Poss}\,(\pi_i)$$

for all π_i and all $i = 1, \ldots, n$.

EXERCISE 10. Show that a game has perfect recall if and only if the following condition is satisfied.

Let X and Y be personal moves for P_i on the same play lying in information sets U and V respectively and let Y follow X by the vth alternative. Let

$$D_v(U) = \left\{Z \mid Z \text{ follows some move in } U \text{ by the } v^{th} \text{ alternative}\right\}.$$

Then it is required that $V \subset D_v(U)$.

[By interpreting this exercise it is seen that games with perfect recall are those in which each player is allowed by the rules to remember everything he knew at previous moves and all of his choices at those moves. This obviates the use of agents; indeed, the only games which do not have perfect recall are those, such as bridge, which include the description of the agents in their verbal rules.

The next two propositions enunciate key properties of games with perfect recall. The first simplifies the calculation of the expected payoffs when playing mixed strategies while the second expresses the "independence" of the plays to work a similar simplification for behavior strategies.]

PROPOSITION 4. Let Γ be a game with perfect recall. For all W and $i = 1, \ldots, n$, let $L_i(W)$ be the set of all π_i which choose the last alternative for P_i on W, if any; otherwise $L_i(W)$ is the set of all π_i. Finally, let $x(W)$ be the product of the chance probabilities on W, if any; otherwise set $c(W) = 1$. Then, for all $\pi = (\pi_1, \ldots, \pi_n)$ and all W:

$$p_W(\pi) = \begin{cases} c(W) & \text{for } \pi_i \in L_i(W) \text{ for } i = 1, \ldots, n \\ 0 & \text{otherwise} \end{cases}$$

PROOF. It is clear that we need only show that $\pi_i \in L_i(W)$ implies $\pi_i(U) = v(e)$ for all alternatives e at $X \in U \subset P_i$ on W. Let the last move Y on W for P_i lie in the information set V. Then

$$\pi_i \in L_i(W) \implies V \in \text{Rel}\,(\pi_i) \implies Y \in \text{Poss}\,(\pi_i)$$

and hence there exist $\pi_1', \ldots, \pi_{i-1}', \pi_{i+1}', \ldots, \pi_n'$ such that $p_Y(\pi_1', \ldots, \pi_i, \ldots, \pi_n')) > 0$. Therefore $\pi_i(U) = v(e)$ for all e at $X \in U \subset P_i$ on W.

PROPOSITION 5.   Let Γ be a game with perfect recall. Let e be an alternative at $X \in U \subset P_i$ on W with next move, Y, for P_i on W, if any such exists, lying in the information set V. Finally, let

$$S = \{\pi_i \mid V \in \text{Rel } (\pi_i)\}$$
$$T = \{\pi_i \mid \pi_i(U) = v(e)\}.$$

Then,

$$S = T.$$

PROOF. Assume $\pi_i \in S$. Then $V \in \text{Rel}(\pi_i) \implies Y \in \text{Poss}(\pi_i) \implies \pi_i(U) = v(e) \implies \pi_i \in T$.

Assume $\pi_i \in T$. Then $U \in \text{Rel }(\pi_i) \implies X \in \text{Poss }(\pi_i) \implies V \in \text{Rel }(\pi_i) \implies \pi_i \in S$.

EXERCISE 11. Show that the example of this section satisfies neither Proposition 4 nor Proposition 5.

[Propositions 4 and 5 are really preparatory lemmas for the principal theorem concerning behavior strategies and games with perfect recall. This says that we can do as well with behavior strategies as with mixed strategies in games with perfect recall; in particular, we can solve zero-sum two-person games with perfect recall by means of behavior strategies. A partial converse to this theorem says that perfect recall is the weakest restriction that we can make on the information to insure this situation; the converse will not be proved here.]

THEOREM 11. Let Γ be a game with perfect recall. Then

$$E_i(\beta(\mu)) = E_i(\mu) \quad \text{for } i = 1, \ldots, n,$$

and all n-tuples of mixed strategies $\mu = (\mu_1, \ldots, \mu_n)$.

PROOF. Clearly it is sufficient to show that

$$p_W(\beta(\mu)) = p_W(\mu) \quad \text{for all } W.$$

If there is an alternative e on W belonging to P_i which is incident at a move in an information set that is irrelevant for μ_i, then both sides are clearly zero. Hence we may assume that all such information sets are relevant. Working with each side separately, we have, first:

$$p_W(\beta(\mu)) = \prod_{e \subset W} p_e(\beta(\mu)).$$

Considering those e on W which belong to P_i, their probabilities are given by the fractions of Definition 20. Taking the product of these fractions, we note that the first denominator is 1 while each numerator is the denominator of the next fraction by Proposition 5. Hence

$$p_W(\beta(\mu)) = c(W) \prod_{i=1}^{n} \left(\sum_{\pi_i \in L_i(W)} q_{\pi_i} \right)$$

where $c(W)$ and $L_i(W)$ are defined as in Proposition 4. On the other hand,

$$p_W(\mu) = \sum_\pi q_{\pi_1} \cdots q_{\pi_n} p_W(\pi).$$

$$= c(W) \sum_\pi q_{\pi_1} \cdots q_{\pi_n}, \quad \text{all } \pi_i \in L_i(W)$$

by Proposition 4. Comparing these expressions we see that the theorem is proved.

COROLLARY. Any zero-sum two-person game with perfect recall can be solved with behavior strategies.

3.8 Simplified Poker Reconsidered

To illustrate the power of Theorem 11 we will give the solution of Simplifed Poker in terms of behavior strategies. The extensive form of this game, after elimination of the dominated choices, is as shown in the following figure:

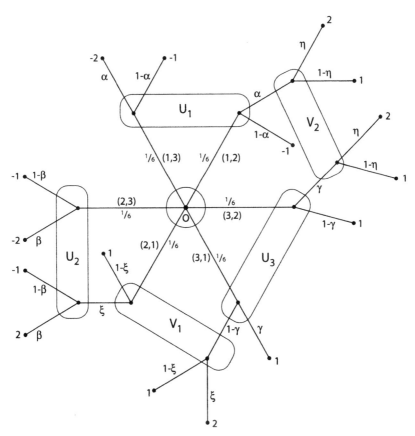

Here, information sets U_1, U_2, U_3 belong to P_1 and if we let α, β, γ be the probability that he plays alternative 2 on these sets, then these correspond to the behavior strategies listed on page 44. Similarly, V_1 and V_2 are information sets for P_2 and we let ξ and η be the probabilities that he plays alternative 2 on these sets. Remark that, since this is a zero-sum two-person game, the numbers labeling the plays are the amounts that P_2 pays to P_1; similarly, we will use the single expected payoff $E((\alpha, \beta, \gamma), (\xi, \eta)) = E_1((\alpha, \beta, \gamma), (\xi, \eta))$. By straightforward calculation,

$$6E((\alpha, \beta, \gamma), (\xi, \eta)) = \alpha - \beta - \xi - 3\alpha\eta + 3\beta\xi - \gamma\xi + \gamma\eta.$$

Then applying customary techniques of the differential calculus to find interior maxima and minima, we have

$$6\frac{\partial E}{\partial \alpha} = 1 - 3\eta \qquad \Longrightarrow \eta = \frac{1}{3}$$

$$6\frac{\partial E}{\partial \beta} = -1 + 3\xi \qquad \Longrightarrow \xi = \frac{1}{3}$$

$$6\frac{\partial E}{\partial \gamma} = -\xi + \eta \qquad \Longrightarrow \xi = \eta$$

$$6\frac{\partial E}{\partial \xi} = -1 + 3\beta - \gamma \Longrightarrow \beta = \frac{\gamma}{3} + \frac{1}{3}$$

$$6\frac{\partial E}{\partial \eta} = -3\alpha + \gamma \qquad \Longrightarrow \alpha = \frac{\gamma}{3}.$$

which coincide with the solution found in Chapter 2.

Notes

1. Since the rules of most games include a "stop rule" which only ensures that every play terminates after a finite number of choices, it is not completely obvious that this entails the finiteness of the game tree. To demonstrate this, following D. König, "Über eine Schlussweise aus dem Endlichen ins Undendliche," *Acta Szeged* **3** (1927), 121–130, we will assume that there are an infinite number of possible plays and then contradict the stop rule by constructing a unicursal path starting at 0 and containing an infinite number of edges. Since the choice at 0 is made from a finite set of alternatives, there must be an infinite number of plays with the same first edge e_1. We proceed by induction; assume that edges $e_1, \ldots, e_\ell$ have been chosen so that $e_1 \ldots e_\ell$ is the beginning segment of an infinite number of plays. Then, since the next choice is made from a finite set, an infinite subset

of these plays must continue with the same edge, say $e_{\ell+1}$. This completes the proof.

2. A partition is an exhaustive decomposition into (possibly void) disjoint sets. Thus we can partition the set $\{1, 2, 3\}$ into the sets $\{1\}$, $\{2, 3\}$, and the void set. To refine a partition is to break it into sets which are no larger than the sets of the given partition. Thus every partition is a refinement of itself.

3. This assumption is weaker than the von Neumann requirement that a player be informed of the number of choices preceeding his move. Thus we would admit the following game, not covered by his formalization.

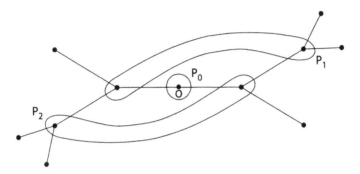

4. See, for example,

Kuhn, H. W., "Extensive games," *Proc. Nat. Acad. Sci.*, **36** (1950), 570–576.

Kuhn, H. W., "Extensive games and the problem of information," in Annals of Math. Study No. 28, Princeton, 1952.

Dalkey, N., "Equivalence of information patterns and essentially determinate games," in Annals of Math. Study No. 28.

5. This reduction is closely related to that of W. D. Krental, J. C. C. McKinsey, and W. V. Quine, "A simplification of games in extensive form," *Duke Math. Jour.*, **18**, (1951), 885–900.

6. Whenever a binary relation, R, is defined on the elements $x, y, z, \ldots$ of a set S satisfying

   (1) $R(x, y) \iff R(y, x)$ (Symmetry)
   (2) $R(x, x)$ for all $x \in S$ (Reflexivity)
   (3) $R(x, y)$ and $R(y, z) \implies R(x, z)$ (Transitivity)

the relation is called an *equivalence relation* and defines a partition of S as follows: the set of the partition which contains an element x consists of all elements y such that $R(x, y)$. The sets of this partition are called *equivalence classes*.

Chapter Four

Infinite Games

4.1 Some Preliminary Restrictions

The games which have been discussed up to now have been restricted by two conditions mentioned in Chapter 1. These are (1) every move has a finite number of alternatives and (2) every play contains a finite number of moves. It is clear that these restrictions can be relaxed in a wide variety of combinations. Actually, only three general types of infinite games have been studied to any extent: the matrix games with a denumerable number or a continuum of pure strategies [1] and infinite game trees in which condition (1) above is satisfied [2]. In this chapter, we will only consider matrix games with a continuum of pure strategies. Among the infinite games, they have been studied the most intensively and, indeed, they seem to present the most cogent interpretations for actual situations. To make precise our exact domain of investigation, we formulate the following definition.

> **DEFINITION 22.** A *zero-sum two-person game on the unit square* is given by any real valued function $A(x, y)$ defined for $0 \leqq x, y \leqq 1$. The values of x constitute the *pure strategies for* P_1 while the values of y are the *pure strategies for* P_2.

The interpretation of games on the unit square is exactly analogous to that of finite matrix games. The game is a two-move game in which P_2 makes his choice of a value y in ignorance of the choice of a value x by P_1. The amount that P_2 then pays to P_1 is $A(x, y)$.

4.2 An Illuminating Example

It is immediately seen that not only are pure strategies insufficient to solve games on the unit square but also finite mixtures of pure strategies will not serve for this purpose. This difficulty is illustrated by the following example.

> **EXAMPLE.** Consider the set of all continuous functions $f(x)$ defined on a closed interval $[0, 1]$. It is a well-known fact that such a function is uniquely specified by giving its values for the rational values of the argument and hence

by familiar cardinality arguments [3] the number of such functions is at most $(2^{\aleph_0})^{\aleph_0} = 2^{\aleph_0 \cdot \aleph_0} = 2^{\aleph_0}$. (Here $\aleph_0$, read "aleph nought", denotes the cardinality of the rational numbers and $2^{\aleph_0}$ denotes the cardinality of the real numbers.) If we restrict ourselves to the functions f which satisfy $|f(x)| \leq 1$ and $\int_0^1 f(x)dx = 0$ this statement remains true. On the other hand there are at least as many such functions as there are real numbers, namely the functions $g_r : g_r(x) = 2rx - r$ where $0 \leq r \leq 1$. Hence there are exactly as many such functions as there are real numbers. Summarizing, there is a one-one correspondence between the real numbers y where $0 \leq y \leq 1$ and the continuous functions f such that $|f(x)| \leq 1$ and $\int_0^1 f(x)dx = 0$. Let us denote the function which corresponds to y by f_y.

We next define a game by setting $A(x, y) = f_y(x)$. I contend that the best that P_1 can assure himself by using a finite mixture is -1 while the best that P_2 can do is break even. To prove the first part, suppose P_1 plays the points $x_1 < x_2 < \cdots < x_m$ with probabilities $p_1, \ldots, p_m$. Then, if P_2 plays his pure strategy y corresponding to the function

$$f(x) = \begin{cases} \dfrac{-2x + x_1}{x_1}, & 0 \leq x \leq x_1 \\[2ex] \dfrac{4x - 3x_i - x_{i+1}}{x_{i+1} - x_i}, & x_i \leq x \leq \dfrac{x_i + x_{i+1}}{2} \\[2ex] \dfrac{-4x + x_i + 3x_{i+1}}{x_{i+1} - x_i}, & \dfrac{x_i + x_{i+1}}{2} \leq x \leq x_{i+1} \\[2ex] \dfrac{2x - x_m - 1}{1 - x_m}, & x_m < x \leq 1. \end{cases} \quad \text{where } i = 1, \ldots, m - 1$$

From the graph of this function,

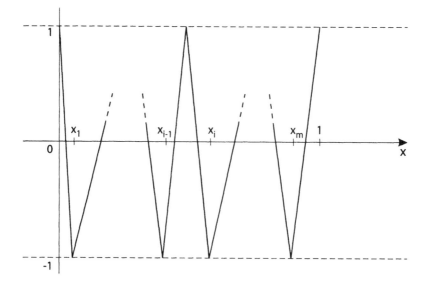

one is easily convinced that $|f(x)| \leqq 1$, $\int_0^1 f(x)dx = 0$ and, what is most important, $f(x_i) = -1$ for $i = 1, \ldots, n$. Since he must take into account this choice of y, P_1 can never assure himself any more than -1 by using a finite mixture of pure strategies.

On the other hand, P_2 can always break even by choosing the y corresponding to the function which is identically 0. To convince ourselves that this is the best that he can do using a finite mixture of the functions available to him, we assume that he plays the functions $f_1, \ldots, f_n$ with probabilities $q_1, \ldots, q_n$. Then his expectation against x is $\sum_j q_j f_j(x)$. If this is negative for all x, then

$$0 > \int_0^1 \sum_j q_j f_j(x) = \sum_j q_j \int_0^1 f_j(x) = 0$$

which is a contradiction.

Now, to the mathematically sophisticated, we could say that the way out of our difficulty is to use measures on the interval to replace our finite probability distributions. However, we do not demand this degree of sophistication and so must spend a short interlude discussing the rudiments of measure theory. However, since the main object is the application to game thoery we shall feel free to leave many statements without proof, giving the reader only an appropriate reference.

4.3 Mixed Strategies and Expectation

The example of the last section makes it plausible that finite mixtures will not suffice for games on the square. If not finite mixtures, then what? Let us proceed by a procedure common in mathematics. We now suspect that our "mixed strategies" must be wider than the finite mixtures; the next step will be to try to narrow the class of possibilities by the desired properties of such a plan.

First of all, we want the intuitive properties of probability. As they apply to the unit interval they might be stated:

1. Given a subset S of $[0, 1]$, the probability that the pure strategy chosen lies in S is a number between zero and one.
2. Given two disjoint subsets S_1 and S_2 of $[0, 1]$, the probability that the pure strategy chosen lies in $S_1 \cup S_2$ is the probability that it lies in S_1 plus the probability that it lies in S_2.
3. The probability that the strategy chosen lies in $[0, 1]$ is one.

Mathematical experience, if not practical experience, has shown that this concept is too broad in two directions [4]. First, we should not (and, in a certain sense, we *cannot* [5]) assign a probability to all subsets of the unit interval. Second, it proves to be convenient to strengthen requirement (2) to include denumerable unions of disjoint sets. The justification of this restriction will be that the narrower class of probability assignments allowed in this manner is still large enough to solve the games that we shall consider. Such functions defined on a suitably restricted class of subsets of [0, 1] are called *probability measures*.

DEFINITION 23. A *probability measure* μ on the interval [0, 1] is a real-valued function defined on a class $\mathcal{B}$ of subsets S of [0, 1] such that:

    (I) $\mu(S) \geq 0$ for $S \in \mathcal{B}$;
    (II) if S is the disjoint union of sets $S_k \in \mathcal{B}, k = 1, 2, \ldots$ then
$$\mu(S) = \sum_i \mu(S_i);$$
    (III) $\mu([0, 1]) = 1.$

The class $\mathcal{B}$ is called the class of *Borel sets* and is defined to be the smallest class of subsets of [0, 1] containing the subintervals and such that:

    (V) if $S_1 \in \mathcal{B}$ and $S_2 \in \mathcal{B}$ then $S_1 - S_2 \in \mathcal{B}$;
    (VI) if $S_k \in \mathcal{B}, k = 1, 2, \ldots$, then $\bigcup_{i=1}^{\infty} S_k \in \mathcal{B}.$

It should be remarked that the concept of a probability measure is non-vacuous since it includes at least the assignments of probabilities $p_1, \ldots, p_n$ to the points $x_1, \ldots, x_n$ such that $p_1 + \cdots + p_n = 1$. Here μ is defined by

$$\mu(S) = \sum_{\{i \,|\, x_i \in S\}} p_i \quad \text{for all } S \in \mathcal{B}$$

EXERCISE 12. Show that μ, thus defined, is a probability measure.

Probability measures, though close to our intuition, are hard to handle with the usual tools of analysis since they are defined not on points but on sets. For this reason we introduce a class of point functions which can be shown to be equivalent.

DEFINITION 24. The *distribution function* (or simply the *distribution*) F corresponding to a probability measure μ is a real-valued function defined for $x, 0 \leq x \leq 1$, by:

$$F(0) = 0$$

$$F(x) = \mu([0, x]) \quad \text{for } 0 < x \leq 1$$

The probability interpretation is clear; for $x > 0$, $F(x)$ is the probability that a number between 0 and x will be chosen. The exception at $x = 0$ is a concession to later technical simplicity.

THEOREM 12. Any distribution F satisfies:

1. $F(x) \geq 0$ for $0 \leq x \leq 1$.
2. $F(0) = 0$ and $F(1) = 1$

3. if $x_1 \leqq x_2$ then $F(x_1) \leqq F(x_2)$ (F is monotonically increasing).
4. F is right-hand continuous except possibly at $x = 0$.

PROOF. Properties (1) and (2) are immediate consequences of (I) and (II) in Definition 23. On remarking that $x_1 \leqq x_2$ implies that $[0, x_2]$ is the disjoint union of $[0, x_1]$ and the left-open interval $(x_1, x_2]$ we have

$$F(x_2) = \mu([0, x_2]) = \mu([0, x_1]) + \mu((x_1, x_2]) \quad \text{by (II)}$$

$$\geqq \mu([0, x_1]) \quad \text{by (I)}$$

$$= F(x_1).$$

Thus only (4) remains to be proven. Here we must show that $\lim_{x \to a+0} F(x) = F(a)$ for all $a > 0$. Consider any decreasing sequence $x_1 > x_2 > \cdots > x_k > \cdots$ which tends to the fixed value a. Letting the left-open intervals $(a, x_k]$ be denoted by $S_k, k = 1, 2, \ldots$, we have

$$S_1 = (S_1 - S_2) \cup (S_2 - S_3) \cup \cdots = (S_1 - S_2) \cup \cdots \cup (S_{k-1} - S_k) \cup S_k$$

where the sets appearing in the union are disjoint. Hence, by (II),

$$\mu(S_1) = \mu(S_1 - S_2) + \mu(S_2 - S_3) + \cdots$$

$$= \mu(S_1 - S_2) + \cdots + \mu(S_{k-1} - S_k) + \mu(S_k).$$

But the convergence of the infinite series which is the center term of this equation implies that we can make the remainder

$$\mu(S_k) = \mu((a, x_k]) = F(x_k) - F(a)$$

as small as we please by choosing k large enough. Hence $\lim_{k \to \infty} F(x_k) = F(a)$ and property (4) is verified.

Since it is the probability measures that lie closer to our intuition it is important to know that the converse of this theorem holds. It may be stated:

THEOREM 13. Given a real-valued function $F(x)$ defined for $0 \leqq x \leqq 1$ and satisfying conditions (1)–(4) of Theorem 1, there exists a probability μ on $\mathcal{B}$ such that:

$$F(x) = \mu([0, x]) \quad \text{for } 0 < x \leqq 1.$$

We will not prove this theorem, although it is at the base of all that follows. It is a problem of such technical complexity to construct probability measures for even the simplest distributions such as $F(x) = x$ that the proof would surely overshadow the main point at issue which is the definition of mixed strategies for games on the square. Many readable proofs are readily accessible in the literature on probability [6].

Suppose that we agree that P_1 will choose as his mixed strategy the distribution $F(x)$. The next problem that confronts us is to assign an expectation against each pure strategy y for P_2. Assuming that such an expectation can be defined, let us denote it by $E(F, y)$ and try to circumscribe it by considerations of plausibility.

Consider any left-open subinterval $(a, b]$ of $[0, 1]$; player P_1 plays some x in this subinterval with probability $\mu((a, b]) = F(b) - F(a)$. If m is any number such that $m \leqq A(x, y)$ for all x in $(a, b]$, that is, if m is a *lower bound* for $A(x, y)$, then certainly P_1 can expect at least $m(F(b) - F(a))$ from the plays which involve an x in $(a, b]$. This remains true if we take m to be the *greatest lower bound* or *infimum* of $A(x, y)$ for x in $(a, b]$. (The word infimum is preferred because it immediately suggests that it is lower than the values under consideration while the first word that strikes us in the alternative, "greatest lower bound," is "greatest"; infimum is abbreviated inf.). On the other hand, if M is the *least upper bound* or *supremum* of $A(x, y)$ on $(a, b]$ then P_1 can expect no more than $M(F(b) - F(a))$ from those plays (x, y) where x is in $(a, b]$ for y still fixed.

This reasoning motivates the following definition.

DEFINITION 25. Given any subdivision $\triangle$ of $[0, 1]$ by $n + 1$ points $0 = x_0 < x_1 < \cdots < x_n = 1$, let

$$m_1 = \inf_{0 \leqq x \leqq x_1} A(x, y), \qquad m_k = \inf_{x_{k-1} < x \leqq x_k} A(x, y), \qquad k = 2, \ldots, n.$$

$$M_1 = \sup_{0 \leqq x \leqq x_1} A(x, y), \qquad M_k = \sup_{x_{k-1} < x \leqq x_k} A(x, y), \qquad k = 2, \ldots, n.$$

Then

$$s_\triangle = \sum_{k=1}^{n} m_k(F(x_k) - F(x_{k-1}))$$

and

$$S_\triangle = \sum_{k=1}^{n} M_k(F(x_k) - F(x_{k-1}))$$

are called the *lower* and *upper Darboux sums* of $A(x, y)$ for the subdivision $\triangle$. The number δ, which is the maximum of the n numbers $x_k - x_{k-1}$ for $k = 1, \ldots, n$, is called the *mesh* of $\triangle$.

Then our previous reasoning tells us that $E(F, y)$, if it exists, must satisfy

$$s_\triangle \leqq E(F, y) \leqq S_\triangle.$$

Since it is reasonable to expect that "finer" subdivisions of $[0, 1]$ will give more accurate bounds for $E(F, y)$ we now define.

DEFINITION 26. Given a game on the square defined by $A(x, y)$ and given a distribution $F(x)$, the *expectation $E(F, y)$ of P_1 when playing F against y for P_2* is the common value of the limits $\lim_{\delta \to 0} s_\Delta$ and $\lim_{\delta \to 0} S_\Delta$ if both exist and are equal, independent of the manner of subdivision. If $E(F, y)$ exists it is called the *Riemann-Stieltjes integral with respect to $F(x)$ on* $[0, 1]$, and we shall write:

$$E(F, y) = \int_0^1 A(x, y) \, dF(x).$$

The reader should remark that the name is quite apt for, if we consider the special distribution $F(x) = x$, our definition is just that of the Riemann integral, while it was Stieltjes who first considered replacing x by a non-decreasing function $F(x)$. The relation to the Riemann integral should guide the reader through whatever is unfamiliar in the following.

Remark that it is clear, for all subdivisions Δ_1 and Δ_2, that

$$s_{\Delta_1} \leqq S_{\Delta_2}.$$

This is easily seen by first remarking that is certainly true if Δ_1 and Δ_2 are the same subdivision and that "refining" a subdivision by adding new points we only increase s and decrease S. Thus, if Δ is the subdivision of $[0, 1]$ by all of the points of Δ_1 and Δ_2, we have

$$s_{\Delta_1} \leqq s_\Delta \leqq S_\Delta \leqq S_{\Delta_2}.$$

Hence, the Stieltjes integral exists if there exists a number I such that, given $\epsilon > 0$, we can choose $\delta_0 > 0$ with the property

(1) $$0 \leqq I - s_\Delta < \epsilon$$
$$ \quad \text{for the mesh of } \Delta < \delta_0.$$
$$0 \leqq S_\Delta - I < \epsilon$$

THEOREM 14. The Stieltjes integral $\int_0^1 A(x, y) dF(x)$ for $A(x, y)$ bounded exists if and only if, given $\epsilon > 0$, there exists $\delta_0 > 0$ such that

(2) $$S_\Delta - s_\Delta < \epsilon \quad \text{whenever the mesh of } \Delta < \delta_0.$$

PROOF. To prove the necessity, choose $\epsilon/2$ in (1) and add the inequalities.

To prove the sufficiency, let $I = \inf_\Delta S_\Delta$. Then $I \leqq S_\Delta$ for all Δ since the infimum is a lower bound; the facts that I is the greatest lower bound and s_Δ is a lower bound imply that $s_\Delta \leqq I$ for all Δ. Hence $s_\Delta \leqq I \leqq S_\Delta$, $S_\Delta - s_\Delta < \epsilon$ for some choice of $\delta_0 > 0$ and (1) is true.

It would be too much to ask that the Stieltjes integral exist for all $A(x, y)$. However, the following theorem carries over from the theory of the Riemann integral.

THEOREM 15. If, for fixed y, $A(x, y)$ is a continuous function of x in $[0, 1]$ then $\int_0^1 A(x, y)\, dF(x)$ exists for all distribution F.

PROOF. Since $A(x, y)$ is continuous in the closed and bounded set $[0, 1]$ it is uniformly continuous there. This means that we can choose $\delta_0 > 0$ such that

$$|x' - x''| < \delta_0 \quad \text{implies} \quad |A(x', y) - A(x'', y)| < \epsilon.$$

Hence, for any subdivision $\triangle$ with mesh $\delta < \delta_0$, $0 \leqq M_k - m_k < \epsilon$ and

$$
\begin{aligned}
S_\triangle - s_\triangle &= \sum_{k=1}^n (M_k - m_k)(F(x_k) - F(x_{k-1})) \\
&< \sum_{k=1}^n \epsilon(F(x_k) - F(x_{k-1})) = \epsilon.
\end{aligned}
$$

To summarize the definitions and results of this section, to play a game $A(x, y)$ on the square, P_1 chooses a distribution $F(x)$ and P_2 chooses a distribution $G(y)$. Then the expectation of P_1 for F against y is

$$E(F, y) = \int_0^1 A(x, y)\, dF(x)$$

while the expectation of P_1 for x against $G(y)$ is

$$E(x, G) = \int_0^1 A(x, y)\, dG(y).$$

If $A(x, y)$ is a continuous function of x for each fixed y, then $E(F, y)$ exists for all F and similarly, if $A(x, y)$ is a continuous function of y for each fixed x, then $E(x, G)$ exists for all G.

4.4 The Battle of the Maxmin versus Supinf

We have seen that if $A(x, y)$ is a continuous function of x for each y then $E(F, y)$ exists for each distribution F and each y. Now, as in the case of finite matrix games, P_1 takes a pessimistic (or reasonable) view in evaluating what F insures him. If $E(F, y)$ is bounded from below then he is certainly assured this as his expectation, indeed he can surely count on the greatest of such lower bounds, i.e., $\inf_y E(F, y)$. A fine or mathematical distinction is made between saying that he can count on this or the $\min_y E(F, y)$; the latter phrase is reserved for the case in which there is some y_0 for which $E(F, y_0) = \inf_y E(F, y)$. Thus the infimum of the set $\{1, 1/2, 1/3, \ldots\}$ is 0 but it has no minimum while the set $\{0, 1/2, 1/3, \ldots\}$ has infimum and minimum equal to zero. However, the next theorem is designed to get us by this quibble.

THEOREM 26. If $A(x, y)$ is a continuous function of (x, y) then $E(F, y) = \int_0^1 A(x, y) \, dF(x)$ exists and is a continuous function of y for all distributions F.

PROOF. (The reader should first remarke the difference between continuity in two variables and continuity in each separately. Thus considering

$$A(x, y) = \frac{xy}{x^2 + y^2},$$

and $A(0, 0) = 0$, we have $\lim_{x \to a} A(x, y) = A(a, y)$ for each y and $\lim_{y \to b} A(x, y) = A(x, b)$ for each x, but there are points (ϵ, ϵ) as close as we please to $(0, 0)$ where $A(\epsilon, \epsilon) = 1/2$.) Continuity in (x, y) at (a, b) means that, given $\epsilon > 0$, there exists a $\delta_0 > 0$ such that $\sqrt{(x - a)^2 + (y - b)^2} < \delta_0$ implies $|A(x, y) - A(a, b)| < \epsilon$. Just as in the case of one variable, continuity on the closed and bounded unit square means uniform continuity, which is to say that we can choose one δ_0 which depends on ϵ but not on (a, b). Then, using this fact and two properties of the Stieltjes integral that the reader can verify for himself in the same manner that they are proved for the Riemann integral:

$$|E(F, y') - E(F, y'')| = \left| \int_0^1 ((A(x, y') - A(x, y'')) \, dF(x) \right|$$

$$\leq \int_0^1 |A(x, y') - A(x, y'')| \, dF(x)$$

$$< \int_0^1 \epsilon \, dF(x) = \epsilon \text{ for } |y' - y''| < \delta_0.$$

Hence $E(F, y)$ is a continuous function of y for all F.

COROLLARY. If $A(x, y)$ is a continuous function of (x, y) then $\min_y E(F, y)$ exists for all F.

PROOF. The continuous function $E(F, y)$ on the closed bounded set $[0, 1]$ is bounded on that set and there is a number $\bar{y}$ where $E(F, \bar{y}) = \inf_y E(F, y)$.

Naturally, P_1's objective is to maximixe the estimate $\min_y E(F, y)$ by choosing a propitious F. This time we must fight the battle of maximum versus supremum. Since $A(x, y)$ is a continuous function on the closed unit square there must be a maximum payoff which is in turn an upper bound for $\min_y E(F, y)$. Take v to be the least of the upper bounds for $\min_y E(F, y)$. That is, let

$$v = \sup_F \min_y E(F, y)$$

which says precisely $\min_y E(F, y) \leq v$ for all F, but, given $\epsilon > 0$, there exists an F such that $\min_y E(F, y) < v + \epsilon$. Suppose we choose the following sequence of ϵ tending to $0 : 1, 1/2, 1/3, \ldots, 1/n, \ldots$ and choose, for each n, a distribution F_n such that

$$v \leqq \min_y E(F_n, y) < v + \frac{1}{n}.$$

Then, clearly, $\lim_{n \to \infty} \min_y E(F_n, y) = v$ and it becomes interesting to ask the following questions:

(1) Does there exist a *distribution* F such that $\lim_{n \to \infty} F_n(x) = F(x)$ for all x?

(2) If such a distribution F exists, does $\min_y E(F, y) = v$?

These questions will be answered effectively in the affirmative by the next two theorems.

THEOREM 27. Given any sequence of distributions $F_1, F_2, \ldots$, there exist a sequence of positive integers $n_1 < n_2 < \cdots$ and a distribution F such that

$$\lim_{k \to \infty} F_{n_k}(x) = F(x)$$

for all points x at which F is continuous [7].

PROOF. If we can succeed in defining F on the rational numbers r it is intuitively obvious that we can extend the definition by the condition that F must be right-hand continuous. To this end, let us enumerate all of the rational numbers in some order $r_1, r_2, \ldots, r_k, \ldots$ [8]. Then we construct an infinite square array of positive integers with the properties indicated in the following:

$$n_1^1 < n_2^1 < \cdots < n_\ell^1 < \cdots, \quad \lim_{\ell \to \infty} F_{n_\ell^1}(r_1) = F(r_1)$$

$$n_1^2 < n_2^2 < \cdots < n_\ell^2 < \cdots, \quad \lim_{\ell \to \infty} F_{n_\ell^2}(r_2) = F(r_2)$$

$$\cdots \quad \cdots \quad \cdots \quad \cdots$$

$$n_1^k < n_2^k < \cdots < n_\ell^k < \cdots, \quad \lim_{\ell \to \infty} F_{n_\ell^k}(r_k) - F(r_k).$$

The diagram is easily begun; since the sequence $F_1(r_1), F_2(r_1), \ldots$ is bounded by 0 and 1 it has a limit point. Call this limit value $F(r_1)$ and choose $n_1^1 < n_2^1 < \cdots$ such that $\lim_{\ell \to \infty} F_{n_\ell^1} = F(r_1)$. Suppose we have constructed the first $k - 1$ rows of the diagram. Then the sequence $F_{n_1^{k-1}}(r_k), F_{n_2^{k-1}}(r_k), \ldots$ is bounded by 0 and 1 and hence we can choose $n_1^k < n_2^k < \cdots$, a subsequence of $n_1^{k-1} < n_2^{k-1} < \cdots$ in such a manner that $F_{n_\ell^k}(r_k)$ tend to a limit. We call this limit $F(r_k)$.

Now set $n_k = n_k^k$ for $k = 1, 2, \ldots$. It is clear that $n_1 < n_2 < \cdots$ and for each k, all but a finite number of integers in this sequence lie in the sequence $n_1^k < n_2^k < \cdots$. Hence

$$\lim_{k \to \infty} F_{n_k}(r) = F(r) \quad \text{for all rational numbers } r.$$

We now extend the definition of F to all x in $[0, 1]$ by defining

$$F(x) = \inf_{r > x} F(r).$$

We now must verify that F, so defined, is a distribution and that $\lim_{k \to \infty} F_{n_k}(x) = F(x)$ at all points of continuity of F.

(1) $F(x) \geq 0$ for $0 \leq x \leq 1$. This follows from the fact that $F_{n_k}(r) \geq 0$ implies $F(r) \geq 0$ for all r and hence 0 is a lower bound for $F(r)$ for $r > x$ and all x. Therefore $F(x)$, being the greatest lower bound of such $F(r)$ must be greater than or equal to 0.

(2) $F(0) = 0$ and $F(1) = 1$. Since 0 and 1 are rational numbers $F(0) = \lim_{k \to \infty} F_{n_k}(0) = 0$ and $F(1) = \lim_{k \to \infty} F_{n_k}(1) = 1$.

(3) If $x_1 \leq x_2$ then $F(x_1) \leq F(x_2)$. Since the definition $F(x) = \inf_{r > x} F(r)$ holds whether x is irrational or not, we have

$$F(x_1) = \inf_{r > x_1} F(r) \leq \inf_{r > x_2} F(r) = F(x_2),$$

since more numbers r are taken into account in reckoning the left-hand infimum than in computing the right-hand infimum.

(4) F is right-hand continuous at all $a > 0$. To show this let $x_1 > x_2 > \cdots > x_n > \cdots$ be a sequence the ending to a. Then

$$\lim_{n \to \infty} F(x_n) = \lim_{n \to \infty} \inf_{r > x_n} F(r) = \inf_{r > a} F(r) = F(a).$$

Now take a to be a point of continuity of $F(x)$. Then we can choose $h > 0$ so that $F(a + h) - F(a - h)$ is less than any given $\epsilon > 0$. Let r' and r'' be rational numbers in the open intervals $(a - h, a)$ and $(a, a + h)$ so that:

(*) $$F(a - h) \leq F(r') \leq F(x) \leq F(r'') \leq F(a + h).$$

Furthermore, for every k, we have

$$F_{n_k}(r') \leq F_{n_k}(x) \leq F_{n_k}(r'').$$

But $\lim_{k \to \infty} F_{n_k}(r') = F(r')$ and $\lim_{k \to \infty} F_{n_k}(r'') = F(r'')$. Hence the limit $\lim_{k \to \infty} F_{n_k}(x)$ exists and lies between $F(r')$ and $F(r'')$. Now let ϵ tend to zero and we see by (*) that $\lim_{k \to \infty} F_{n_k}(x) = F(x)$ as was to be shown.

THEOREM 28. Let g be a continuous function on $[0, 1]$ and let F_n be a sequence of distributions which converges to the distribution F at all points of continuity of F. Then

$$\lim_{n \to \infty} \int_0^1 g(x) \, dF_n(x) = \int_0^1 g(x) \, dF(x).$$

PROOF. This theorem is easily verified by a straightforward examination of the Darboux sums defining the integrals involved once it has been remarked that a distribution can have at most a denumerable number of discontinuities [9].

Combining Theorems 27 and 28 for the sequence of distributions F_n considered previously, we have the existence of a subsequence F_{n_k} which converges to a distribution $\bar{F}$ with the property:

$$\lim_{k \to \infty} E(F_{n_k}, y) = E(\bar{F}, y) \quad \text{for all } y.$$

If we let y_0 be such that $E(\bar{F}, y_0) = \min_y E(\bar{F}, y)$ then

$$v \geqq \min_y E(\bar{F}, y) = E(\bar{F}, y_0) = \lim_{k \to \infty} E(F_{n_k}, y_0)$$

$$\geqq \lim_{k \to \infty} \min_y E(F_{n_k}, y) = v$$

and hence

$$v = \min_y E(\bar{F}, y).$$

We can summarize the work of this section by asserting

$$v = \max_F \min_y E(F, y).$$

4.5 The Fundamental Theorem

THEOREM 29. Given any game $A(x, y)$ on the unit square with A continuous in (x, y), there exist distributions $\bar{F}$ and $\bar{G}$ and a real number v such that

$$(1) \quad E(\bar{F}, y) \;=\; \int_0^1 A(x, y)\, d\bar{F}(x) \geqq v \quad \text{for all } y$$

$$(2) \quad E(x, \bar{G}) \;=\; \int_0^1 A(x, y)\, d\bar{G}(y) \leqq v \quad \text{for all } x.$$

PROOF. The discussion of Section 4 yields (without further proof) a distribution $\bar{F}$ such that

$$E(\bar{F}, y) \;=\; \max_F \min_y E(F, y) \text{ for all } y.$$

Thus, if we set $v = \max_F \min_y E(F, y)$, it is clear that (1) is satisfied for this choice of $\bar{F}$ and v and it is certain that this is the best that P_1 can do.

Our problem then is to construct an optimal distribution $\bar{G}$ for P_2. Following Ville [10], we will do this by considering a sequence of *finite* matrix games Γ_n which "approximate" the given game on the unit square. The matrix $A^n = (a_{ij}^n)$

of the nth game in this sequence is constructed in quite a straightforward manner; we cover the unit square with a grid of parallel lines $1/n$ apart and use the values of $A(x, y)$ on the points of intersection as the entries in A^n. Thus

$$a_{ij}^n = A\left(\frac{i}{n}, \frac{j}{n}\right) \quad \text{for } i, j = 1, \dots, n.$$

LEMMA. Let v_n be the value of Γ_n. Then, given $\epsilon > 0$, there exists $N(\epsilon)$ such that

$$v_n < v + \epsilon \quad \text{for } n \geq N,$$

that is,

$$\limsup_{n \to \infty} v_n \leq v.$$

PROOF. Given $\epsilon > 0$, we use the uniform continuity of A to choose $N(\epsilon)$ such that for $n \geq N$

$$|x' - x''| < \frac{1}{n} \text{ and } |y' - y''| < \frac{1}{n} \implies |A(x', y') - A(x'', y'')| < \epsilon.$$

Hence, for each y, if we choose j so that $(j - 1)/n < y \leq j/n$ we have:

$$
\begin{aligned}
\left| E(F, y) - E\left(F, \frac{j}{n}\right) \right| &= \left| \int_0^1 A(x, y)\, dF - \int_0^1 A\left(x, \frac{j}{n}\right) dF \right| \\
&\leq \int_0^1 \left| A(x, y) - A\left(x, \frac{j}{n}\right) \right| dF \\
&< \int_0^1 \epsilon\, dF = \epsilon
\end{aligned}
$$

for all distributions F. On the other hand, $\min_y E(F, y) \leq v$ for all F and hence for each F there exists a y such that $E(F, y) \leq v$. Combining these two statements, for every distribution F, there exists a j such that

$$E\left(F, \frac{j}{n}\right) < v + \epsilon.$$

If we restrict F to the mixed strategies for P_1 in Γ_n, we see that this says that P_1 cannot insure himself $v + \epsilon$ in Γ_n. Hence

$$v_n < v + \epsilon$$

and the lemma is proved.

By the fundamental theorem for matrix games, P_2 has an optimal strategy $\bar{Y}_n$ in each Γ_n. Using Exercise 11 and Definition 24 we can construct a corresponding distribution G_n with the property

$$E\left(\frac{i}{n}, G_n\right) \leq v_n \quad \text{for } i = 1, \dots, n.$$

Again by the continuity of A and the choice of $N(\epsilon)$, for arbitrary $\epsilon > 0$, we have

$$E(x, G_n) < v_n + \epsilon \quad \text{for all } x \text{ and } n \geq N(\epsilon).$$

But Theorem 27 assures the existence of a subsequence G_{n_k} and a distribution $\bar{G}$ such that $G_{n_k} \longrightarrow \bar{G}$ at all points of continuity of $\bar{G}$. Hence, by Theorems 26 and 28,

$$E(x, \bar{G}) = \lim_{k \longrightarrow \infty} E(x, G_{n_k}) < v + 2\epsilon \quad \text{for all } x.$$

But ϵ is arbitrary in this inequality and hence $\bar{G}$ satisfies (2). This completes the proof of the theorem.

It should be remarked that, after the theorem has been proved, symmetry arguments enable us to sharpen the lemma to $\lim_{n \longrightarrow \infty} v_n = v$, which shows that the games Γ_n truly approximate the given game on the square.

4.6 The Solution of Games on the Unit Square

Implicit in the contention that Theorem 29 is the fundamental theorem for games on the unit square with a continuous payoff function was the following definition:

DEFINITION 27. A *solution* of the game on the unit square with payoff function $A(x, y)$ is a pair of distributions $\bar{F}$ and $\bar{G}$ and a real number v such that $E(\bar{F}, y)$ exists for all y, $E(x, \bar{G})$ exists for all x,

$$E(\bar{F}, y) \geq v \quad \text{for all } y,$$

and

$$E(x, \bar{G}) \leq v \quad \text{for all } x.$$

The distributions $\bar{F}$ and $\bar{G}$ are then called *optimal strategies* and v is called the *value* of the game.

It is clear that the Saddlepoint and Minimax statements of Theorem 29 can be formulated and are valid. However, it should not be inferred that games on the square with *continuous* payoff are the only infinite games which have solutions. The fact that this is the simplest sufficient condition must be considered a mathematical accident and, indeed, the majority of "natural" games on the unit square do not satisfy it. It should also be noted that the theorem provides no effective method for solving even this quite restricted class of games. We shall attempt to give something of a taste of the methods used to solve games on the square through the medium of examples.

A fertile source of examples of infinite games is provided by models of common card games. Indeed, they invite treatment by a continuous variable on two grounds. First, the combinatorial complexity of finite models precludes consideration of any but the simplest cases. Second, the natural linear ordering of a large number of hands in games such as Poker virtually invites the passage to a continuum of hands. Consider, for example, the following model of Poker due to Borel [11].

LA RELANCE. An ante of a units is required by each of the two players. At the beginning of a play they receive fixed *hands*, s and t, chosen at random from the unit intervals $0 \leq s \leq 1$ and $0 \leq t \leq 1$ [12]. Then P_1 either bets an amount $b - a$ or drops out, losing his ante. If P_1 bets, then P_2 can either see the bet or drop out, losing his ante. If P_2 sees a bet, the hands are compared with the higher card winning the total wager b.

We shall asume that P_1 uses pure strategies of the following form: he chooses a number x in the unit interval, $0 \leq x \leq 1$, and decides to bet when his hand exceeds x and to drop out otherwise. Correspondingly, a pure strategy for P_2 will consist of the choice of a number y in the unit interval, $0 \leq y \leq 1$, and the decision to see any possible bet when his hand exceeds y and to drop out otherwise. (The careful reader will want to verify that every other pure strategy is dominated by one of these. He will also remark that the original game had too many strategies to be a game on the square). Then the expected payoff to P_1 is easily computed to be

$$A(x, y) = -ax + a(1 - x) y + \begin{cases} b(1 - y)(x - y) & \text{for } x < y \\ b(1 - x)(x - y) & \text{for } x > y \end{cases}$$

by examination of the following figures.

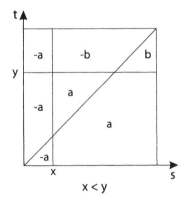

$x < y$

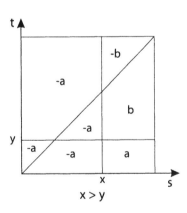

$x > y$

These figures do *not* represent the unit square of strategies: they are the unit squares of possible deals and the payoffs shown depend on the relative size of x and y.

To develop a feeling for the cross-sections of the payoff surface, we can rearrange the payoff functions as polynomials in x and y as follows:

$$x < y : A(x, y) = \begin{cases} [ay - by(1 - y)] + [(b - a) - (a + b)y]x \\ (b - a)x - [(b - a) + (a + b)x]y + by^2 \end{cases}$$

$$x = y: \ A(x, y) = -ax^2 = -ay^2$$

$$x > y: \ A(x, y) = \begin{cases} -(b - a)y + (b - a)(1 + y)x - bx^2 \\[2mm] [-ax + bx(1 - x)] - (b - a)(1 - x)y. \end{cases}$$

From these formulas it is easily seen that the cross-sections of the surface consist of a piece of a parabola continued by a line segment. They are graphed in the following diagrams:

Cross-sections for fixed x

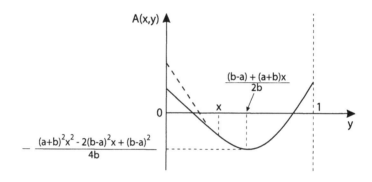

Here it is easily verified that the line segment to the left of x is tangent to the segment of the parabola to the right of x. The line segment has the non-positive slope $-b(b - a)(1 - x)$ and hence the minimum always occurs to the right of x as shown.

Cross-sections for fixed y

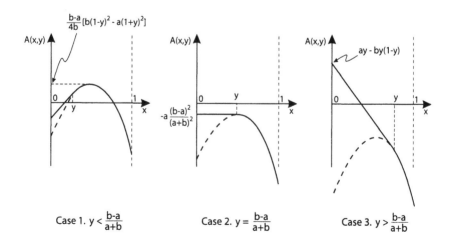

Case 1. $y < \dfrac{b-a}{a+b}$ 　　　　Case 2. $y = \dfrac{b-a}{a+b}$ 　　　　Case 3. $y > \dfrac{b-a}{a+b}$

Here the three cases are distinguished by the sign of the slope of the line segment to the left of y; again the position of the maximum is established by noting that in all cases the line is tangent to the parabola.

From the cross-sections for fixed x, we compute

$$\min_{y} A(x, y) = -[(a + b)^2 x^2 - 2(b - a)^2 x + (b - a)^2]/4b$$

and hence,

$$\max_{x} \min_{y} A(x, y) = -a \left(\frac{b - a}{a + b} \right)^2$$

the maximum being attained at

$$x = \left(\frac{b - a}{a + b} \right)^2 .$$

We can bypass the tedious but straightforward calculation of $\min_y \max_x A(x, y)$ from the cross-sections with y fixed by noting that

$$\max_x A(x, (b - a)/(a + b)) = -a \left(\frac{b - a}{a + b} \right)^2$$

$$\min_{y} \max_{x} A(x, y) \leqq -a \left(\frac{b - a}{a + b} \right)^2 = \max_{x} \min_{y} A(x, y) .$$

We have seen previously (page 23) that always

$$\min_{y} \max_{x} A(x, y) \geqq \max_{x} \min_{y} A(x, y)$$

and hence for this game

$$\min_{y} \max_{x} A(x, y) = \max_{x} \min_{y} A(x, Y) = -a \left(\frac{b - a}{a + b} \right)^2 .$$

These equations state that La Relance has a solution in pure strategies with the value

$$v = -a \left(\frac{b - a}{a + b} \right)^2 .$$

The saddlepoint on the payoff surface can be seen by comparing the following figure with the cross-sections given above.

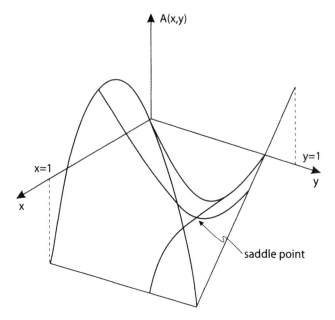

If we want to give the optimal pure strategies as distributions by the use of Exercise 11 and Definition 24, it is convenient to introduce the notation $I_z(x)$ for the "one-step" distribution:

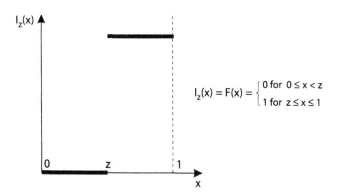

$$I_z(x) = F(x) = \begin{cases} 0 \text{ for } 0 \le x < z \\ 1 \text{ for } z \le x \le 1 \end{cases}$$

Then the optimal pure strategy for P_1 becomes

$$I_{((b-a)/(a+b))^2}$$

and the optimal pure strategy for P_2 is

$$I_{((b-a)/(a+b))} \cdot$$

EXERCISE 13. Show that the optimal strategy given for P_2 is unique, while P_1 can use any distribution $\bar{F}$ satisfying:

$$\int_0^1 x\bar{F}(x)\,dx = \left(\frac{b-a}{a+b}\right)^2$$

$$\bar{F}(x) = 1 \quad \text{for } \frac{b-a}{a+b} \leqq x \leqq 1.$$

Interpret this result in Poker terms.

It is no accident that this game possesses a solution in pure strategies; that is a direct consequence of the geometrical configuration of the cross-sections [13]. These are convex functions of y and concave functions of x, respectively, as defined by:

DEFINITION 28. A real-valued function f defined for y in an interval $a \leq y \leq b$ is said to be *convex* in the interval if

$$\lambda f(y_1) + \mu f(y_2) \geqq f(\lambda y_1 + \mu y_2)$$

for all $\lambda > 0$, $\mu > 0$, $\lambda + \mu = 1$ and $y_1 \neq y_2$ in the interval $a \leq y \leq b$. If $>$ is always valid, f is said to be *strictly convex*. A function f is called *concave* (*strictly concave*) if $-f$ is convex (strictly convex).

EXERCISE 14. (a) Show that the sum of two convex functions is convex.
(b) Show that, if f is a continuous, strictly convex function of y, then there is a unique $\bar{y}$ where $\min_y f(y)$ is achieved and that, for all $\epsilon > 0$, there exists a $\delta > 0$ with

$$f(y) > f(\bar{y}) + \delta \text{ for } |y - \bar{y}| > \epsilon .$$

With these definitions we are in a position to "explain" by a general theorem the existence of a unique pure optimal strategy for P_2 in La Relance.

THEOREM 30. Given any game $A(x, y)$ on the unit square with A continuous in (x, y) and such that $A(x, y)$ is a strictly convex function of y for each x, then there exists a unique optimal pure strategy $\bar{y}$ for P_2.

PROOF. The existence of a solution to the game is guaranteed by Theorem 29. Let v be the value of the game and let $\bar{F}$ be an optimal strategy for P_1. Theorem 26 asserts that $E(\bar{F}, y)$ is a continuous function of y while

$$\lambda E(\bar{F}, y_1) + \mu E(\bar{F}, y_2) \;=\; \lambda \int_0^1 A(x, y_1)\, d\bar{F}(x)$$

$$+ \mu \int_0^1 A(x, y_2)\, d\bar{F}(x)$$

$$= \int_0^1 (\lambda A(x, y_1) + \mu A(x, y_2))\, d\bar{F}(x)$$

$$> \int_0^1 A(x, \lambda y_1 + \mu y_2)\, d\bar{F}(x)$$

$$= E(\bar{F}, \lambda y_1 + \mu y_2)$$

for $\lambda > 0$, $\mu > 0$, $\lambda + \mu = 1$, and $y_1 \neq y_2$, and hence $E(\bar{F}, y)$ is a strictly convex function of y. Therefore, by Exercise 14 there exists a unique $\bar{y}$ such that, for all $\epsilon > 0$, there exists a $\delta > 0$ with

$$E(\bar{F}, y) > E(\bar{F}, \bar{y}) + \delta \text{ for } |y - \bar{y}| > \epsilon.$$

If $\bar{G}$ is any optimal strategy for P_2, let $p(\epsilon)$ denote the probability assigned by $\bar{G}$ to pure strategies y with $|y - \bar{y}| > \epsilon$ (that is, the probability

$$\int_0^{\bar{y}-\epsilon} d\bar{G}(y) + \int_{y+\epsilon}^1 d\bar{G}(y)).$$

Then

$$E(\bar{F}, \bar{y}) = v = \int_0^1 A(x, y)\, d\bar{F}(x)\, d\bar{G}(y) = \int_0^1 E(\bar{F}, y)\, d\bar{G}(y)$$

$$\geqq E(\bar{F}, \bar{y}) + \delta\, p(\epsilon).$$

Hence $p(\epsilon) = 0$ for all $\epsilon > 0$ and $\bar{G}$ assigns probability one to $\bar{y}$.

Theorem 30 is somewhat unsatisfactory in that it presupposes the general existence theorem and deals with distributions in its proof while it is obvious that the result is a statement about pure strategies. We shall remove these objections by a two-sided theorem that also applies to La Relance and which, incidentally, gives an independent proof of the existence theorem for the type of games that it handles.

LEMMA 1. Let f and g be continuous convex functions defined for $a \leqq y \leqq b$ such that either $f(y) > 0$ or $g(y) > 0$ for each such y. Then there exist $\lambda \geqq 0$ and $\mu \geqq 0$ with $\lambda + \mu = 1$ such that

$$\lambda f(y) + \mu g(y) > 0 \quad \text{for all } y, a \leqq y \leqq b.$$

PROOF. Let A denote the set of y for which $f(y) \leqq 0$ and let B denote the set of y for which $g(y) \leqq 0$. Clearly, $A \cap B = \emptyset$ and $\lambda f(y) + \mu g(y) > 0$ for any choice of λ and μ if y lies neither in A nor in B. If A is void then $\lambda = 1$ and $\mu = 0$ satisfy the lemma, while if B is void then $\lambda = 0$ and $\mu = 1$ is a satisfactory choice. Otherwise, let

$$\min_{y \in A} \frac{f(y)}{g(y)} = \frac{f(p)}{g(p)} \quad \text{and} \quad \min_{y \in B} \frac{g(y)}{f(y)} = \frac{g(q)}{f(q)}$$

and suppose that we can find λ and μ such that

(1) $\lambda f(p) + \mu g(p) > 0 \quad \text{and} \quad \lambda f(q) + \mu g(q) > 0.$

Under these conditions, y in A implies

$$\frac{f(y)}{g(y)} \geqq \frac{f(p)}{g(p)} \quad \text{with } g(y) > 0 \text{ and } g(p) > 0,$$

hence

$$\lambda f(y) \geqq \lambda \frac{f(p)}{g(p)} g(y) > -\mu g(p) \frac{g(y)}{g(p)}$$

and therefore

$$\lambda f(y) + \mu g(y) > 0.$$

Similarly,

$$\lambda f(y) + \mu g(y) > 0 \quad \text{for all } y \in B.$$

To assert the existence of a λ and μ satisfying (1) above is equivalent to stating that the 2×2 matrix game

$$\begin{pmatrix} f(p) & f(q) \\ g(p) & g(q) \end{pmatrix}$$

has a positive value. If this is not the case, then there exist (by Theorem 1) $\alpha \geqq 0$ and $\beta \geqq 0$ with $\alpha + \beta = 1$ and such that

$$\alpha f(p) + \beta f(q) \leqq 0 \quad \text{and} \quad \alpha g(p) + \beta g(q) \leqq 0.$$

However, by the convexity of f and g,

$$f(\alpha p + \beta q) \leqq \alpha f(p) + \beta f(q) \quad \text{and} \quad g(\alpha p + \beta q) \leqq \alpha g(p) + \beta g(q)$$

and hence we would have

$$f(\alpha p + \beta q) \leqq 0 \quad \text{and} \quad g(\alpha p + \beta q) \leqq 0$$

contrary to the assumption of the lemma. Therefore the λ and μ satisfying (1) exist and the lemma is proved.

LEMMA 2. Let $f_1, \ldots, f_m$ be continuous convex functions defined for $a \leqq y \leqq b$ such that some $f_i(y) > 0$ for each such y. Then there exist $\lambda_1 \geqq 0, \ldots, \lambda_m \geqq 0$ with $\lambda_1 + \cdots + \lambda_m = 1$ such that

$$\lambda_1 f_1(y) + \cdots + \lambda_m f_m(y) > 0$$

for all $y, a \leqq y \leqq b$.

PROOF. The assertion is a tautology for $m = 1$ and is confirmed by Lemma 1 for $m = 2$. Assume that it has been shown to be true for fewer than m functions and let S denote the set of y for which $f_m(y) \leqq 0$. If S is void then the values $\lambda_1 = \cdots = \lambda_{m-1} = 0$ and $\lambda_m = 1$ satisfy the lemma. Otherwise, since f_m is convex and continuous, S is a closed interval of values of y and we can apply the induction hypothesis to $f_1, \ldots, f_{m-1}$ on this interval to obtain

$$g(y) = \lambda_1' f_1(y) + \cdots + \lambda_{m-1}' f_{m-1}(y)$$

with $\lambda_1' \geqq 0, \ldots, \lambda_{m-1}' \geqq 0, \lambda_1' + \cdots + \lambda_{m-1}' = 1$, and $g(y) > 0$ for all $y \in S$. Then we apply Lemma 1 to the functions f_m and g to obtain

$$\lambda f_m(y) + \mu g(y) > 0 \quad \text{for all } y, a \leqq y \leqq b.$$

That is, the values $\lambda_1 = \mu \lambda_1', \ldots, \lambda_{m-1} = \mu \lambda_{m-1}', \lambda_m = \lambda$ satisfy the present lemma and the proof is complete.

THEOREM 31. Given any game $A(x, y)$ on the unit square with A continuous in (x, y) and such that $A(x, y)$ is a concave function of x for each y and a convex function of y for each x, then there exists a solution in *pure* strategies, $\bar{x}, \bar{y}$ and v.

PROOF. Let $v = \max_x \min_y A(x, y)$ and let $\bar{x}$ be the pure strategy that achieves this maximum assured payoff for P_1. With this definition of v, either there is a $\bar{y}$ such that $A(x, \bar{y}) \leqq v$ for all x and we have the desired solution or, for each y, there is an $x = \varphi(y)$ such that

$$A(\varphi(y), y) > v.$$

If we let O_X denote the set of those y such that $A(x, y) > v$, we have $y \in O_{\varphi(y)}$ and hence the *open* sets O_x cover $0 \leqq y \leqq 1$. By the compactness of the unit interval [14] we can choose a finite number of these sets, $O_{x_1}, \ldots, O_{x_m}$, such that every y lies in one of these sets. Then, if we define $f_i(y) = A(x_i, y) - v$ for $i = 1, \ldots, m$, it is clear that these are continuous convex functions such that some $f_i(y) > 0$ for all y. Therefore, by Lemma 2, there exist $\lambda_1 \geqq 0, \ldots, \lambda_m \geqq 0, \lambda_1 + \cdots + \lambda_m = 1$ such that

$$\lambda_1 f_1(y) + \cdots + \lambda_m f_m(y) > v \quad \text{for all } y.$$

By the concavity of $A(x, y)$ in x, if we set $x_0 = \lambda_1 x_1 + \cdots + \lambda_m x_m$,

$$A(x_0, y) \geq \sum_i \lambda_i A(x_i, y) = \sum_i \lambda_i f_i(y) > v \quad \text{for all } y.$$

However, this contradicts the definition of v and proves the theorem.

Notes

1. See, for example, the pioneering papers:

Ville, J., "Sur la théorie générale des jeux où intervient l'habilité des joueurs," *Traité du calcul des probabilités et de ses applications*, by E. Borel and others, Paris (1938), Vol. 2, No. 5, 105–113.

Wald, A., "Generalization of a theorem by von Neumann concerning zero-sum two-person games," *Annals of Math.*, **46** (1945), 281–286.

Other references may be found in the bibliographies of Annals of Math. Studies Nos. 24 and 28, *Contributions to the Theory of Games*, Vols 1, 2, Princeton, 1950.

2. The study of such games was initiated by D. Gale, F. M. Stewart, "Infinite games with perfect information," in Annals of Math. Study No. 28, Princeton, 1953.

3. The reader unfamiliar with these arguments will find the chapter in G. Birkhoff, S. MacLane, *A Survey of Modern Algebra*, Macmillan, New York, 1941, concerned with transfinite arithmetic (Chapter XII) sufficient for the purposes of these assertions.

4. A good account of the heuristic connections between probability theory and measure theory can be found in pp. 184–191 of: P. R. Halmos, *Measure Theory*, Van Nostrand, New York, 1950.

5. See, for instance, Theorem D, page 69, of Halmos.

6. In addition to Halmos, which may be too detailed for the beginner, one might recommend Chapter 6 of: H. Cramer, *Mathematical Methods of Statistics*, Princeton, 1946.

7. This theorem is sometimes called the Helly Selection Principle; we have followed the proof of Cramer in the reference above.

8. See, for instance, Birkhoff and MacLane, Note 3, above.

9. A distribution can only have "jumps" as discontinuities. Since it is monotone with values between 0 and 1, there can be at most two jumps of amount greater

than or equal to 1/2, at most three for 1/3, and in general, at most n jumps of amount greater than or equal to $1/n$. This enumeration catches all discontinuities.

10. See Ville, Note 1, above.

11. This example is found in Ville.

12. The phrase "chosen at random" is an abbreviation for "chosen according to the uniform probability $F(x) = x$".

13. Games on the square with convex payoff were first studied by F. Bohnenblust, S. Karlin, and L. S. Shapley, "Games with continuous convex payoff," in Annals of Math. Studies No. 24, Princeton, 1950. Theorem 30 can be found in this paper. Theorem 31 is a reworking of methods of Ville and H. Kneser, "Sur un théorème fondamental de la théorie des jeux," *Comptes Rendus*, (June, 1952).

14. The property referred to here is that established by the Heine-Borel Theorem. See, for example, D. V. Widder, *Advanced Calculus*, Prentice-Hall, New York, 1947, pp. 145–146.

Index